A Qualitative Analysis Supplement

SEVENTH EDITION

Kenneth W. Whitten
University of Georgia, Athens

Raymond E. Davis
University of Texas at Austin

M. Larry Peck
Texas A&M University

George G. Stanley
Louisiana State University

THOMSON

BROOKS/COLE

Australia • Canada • Mexico • Singapore • Spain
United Kingdom • United States

We dedicate this book to the memory of our longtime editor, publisher, mentor, and friend, John Vondeling (1933–2001)

Chemistry Editor: John Holdcroft
Development Editor: Jay Campbell
Assistant Editor: Karoliina Tuovinen
Editorial Assistant: Lauren Raike
Technology Project Manager: Ericka Yeoman-Saler
Executive Marketing Manager: Julie Conover
Marketing Assistant: Melanie Wagner
Advertising Project Manager: Stacey Purviance
Project Manager, Editorial Production: Lisa Weber
Print Buyer: Jessica Reed
Permissions Editor: Elizabeth Zuber

Production Service: Sparkpoint Communications
Text Designer: Preston Thomas
Photo Researcher: Dena Digilio Betz
Copy Editor: Sara Bernhardt Black
Illustrators: J/B Woolsey Associates, Rolin Graphics, Inc.
Cover Designer: Joe Fierst and Denise Davidson
Title Page Image: Courtesy of George Stanley
Cover Printer: Phoenix Color Corp
Compositor: G & S Typesetters, Inc.
Printer: Transcontinental Printing/Interglobe

Library of Congress Control Number: 2003104730

A Qualitative Analysis Supplement: ISBN 0-534-40876-1

Brooks/Cole—Thomson Learning
10 Davis Drive
Belmont, CA 94002
USA

Asia
Thomson Learning
5 Shenton Way #01-01
UIC Building
Singapore 068808

Australia/New Zealand
Thomson Learning
102 Dodds Street
Southbank, Victoria 3006
Australia

Canada
Nelson
1120 Birchmount Road
Toronto, Ontario M1K 5G4
Canada

Europe/Middle East/Africa
Thomson Learning
High Holborn House
50/51 Bedford Row
London WC1R 4LR
United Kingdom

Latin America
Thomson Learning
Seneca, 53
Colonia Polanco
11560 Mexico D.F.
Mexico

Spain/Portugal
Paraninfo
Calle/Magallanes, 25
28015 Madrid, Spain

CONTENTS OVERVIEW

CONTENTS

Qualitative Analysis

Some precipitates observed in qualitative analysis. Clockwise from the top: $MgNH_4AsO_4$ (white), CuS (black), $Ni(HDMG)_2$ (bright red), Sb_2S_3 (orange-red), CdS (yellow), and MnS (salmon). Some oxidation of MnS occurred under the photographer's hot lights.

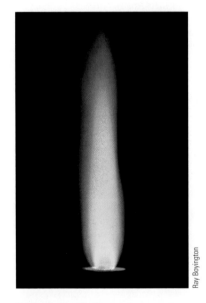

Salts of the alkali metals impart characteristic colors to flames. Sodium is the most distinctive with a bright yellow flame.

Qualitative analysis involves the use of simple chemical tests to identify certain elements present in compounds or mixtures of compounds. These tests typically involve the cationic forms of the common alkali, alkaline earth, transition metals, and a variety of main group metalloids. There are also chemical tests available to identify the common anions (monoatomic and polyatomic) of the more electronegative main group elements, but this book will concentrate on identification procedures for a variety of common metal and metalloid cations. The adjective "qualitative" refers to the fact that these tests usually identify certain metal cations only when they are present in reasonably high concentrations (0.005 M or higher) and do not give accurate information on the exact amounts of each element present.

You may wonder why one would be interested in using these types of simple qualitative chemical tests to identify the presence of some common metallic elements. There are, after all, a wide variety of sophisticated analytical instruments that can identify virtually every element (or compound) accurately down to concentrations of one part per trillion (or lower). But these instruments typically are quite large (desk-sized), expensive ($50,000 to $1,000,000), and certainly not rugged. If you are a geologist in the Rocky Mountains interested in identifying whether a rock or ore sample has iron and/or copper in it, the qualitative analysis tests in this book would allow you to answer that question in 30 minutes with some simple supplies that cost about $30 and can easily be carried in part of a small backpack. No power or complicated (and expensive) electronic equipment is required! With a little more time, you could identify the possible presence of 17 other common metallic or metalloid elements discussed in this book, *if* they are present in high enough concentration. Although not discussed here, there are relatively simple qualitative analysis tests for a large majority of the metals and metalloids on the periodic table.

Another reason for studying qualitative analysis is that it provides students with an excellent introduction to a variety of important laboratory techniques and strategies. Quite a bit of interesting reaction chemistry is involved in these tests (some of it very colorful), and it fits in well with the material that you are learning (or have learned) in your general chemistry classes, especially regarding equilibrium, acid-base, and coordination chemistry concepts. You will find that identifying each component in an unknown mixture of cations is quite a challenging exercise in logic. Keeping a good laboratory notebook and thinking logically about the proper procedure for identifying the cations in an unknown mixture is good training for anyone interested in a science career. It also helps to develop skills in analytical thinking.

This book is a supplement to *General Chemistry*, Seventh Edition, by Whitten, Davis, Peck, and Stanley, which has 28 chapters. Eight additional chapters are included in this supplement. Because it was previously printed together with the main textbook, the first chapter in this supplement is Chapter 29, and the page numbering starts where the last page in the main textbook left off. Many of the references here refer to the main textbook, but this supplement can be used independently. Equivalent topics, which are here referenced to the main textbook, can be found in almost any general chemistry book, though in different specific locations.

The 20 cations (19 elements and NH_4^+) discussed in this book are broken down into five analytical groups based on the chemical properties of their compounds that allow them to be separated from one another and identified. These groupings do *not* correlate with the vertical groups (columns) of elements in the periodic table. When we talk about Group I from the periodic table, we are referring to the alkali metals (and hydrogen). But for our qualitative analysis discussions, analytical Group I refers to Cu, Cd, and Bi. Also note that different qualitative analysis books may define their groupings of element cations differently from what is done here.

In Chapter 29, important properties of the metals of the cation groups are tabulated and discussed, the sources of the elements listed, their metallurgies described, and a few uses of each metal given. Chapter 30 contains a detailed introduction to the laboratory procedures used in semimicro qualitative analysis.

Chapters 31 through 35 cover the analysis of the groups of common cations. Each chapter includes a discussion of the important oxidation states of the metals and metalloids, an introduction to the analytical procedures, and comprehensive discussions of the chemistry of each cation group. Detailed laboratory instructions, set off in color, are also included. These instructions alert you in advance to pitfalls and possible errors, and alternate confirmatory tests and "clean-up" procedures are described for troublesome cations. A set of exercises accompanies each chapter.

In this supplement, we no longer include the cations that create serious disposal or health problems, so mercury, silver, lead, and most chromium reactions have been removed. For this reason, the analysis groups are numbered here somewhat differently from the numbering you might find in many older qualitative analysis schemes.

Chapter 31 describes the analysis of Group I, the metal cations whose sulfides are insoluble in acid solution—bismuth, copper, and cadmium. Cation Group II, described in Chapter 32, includes cations whose sulfides are soluble in basic solution—antimony, tin, and arsenic. Chapter 33 covers the analysis of cations whose sulfides or hydroxides are insoluble in bases; the cations of cobalt, nickel, iron, manganese, aluminum, chromium, and zinc constitute analytical Group III. The insoluble carbonates of the Group IV metals —barium, strontium, and calcium—are the subject of Chapter 34. Chapter 35 discusses Group V, the soluble cations—magnesium, sodium, potassium, and ammonium. Finally, Chapter 36 contains a discussion of some of the more detailed aspects of ionic equilibria as they relate to qualitative analysis.

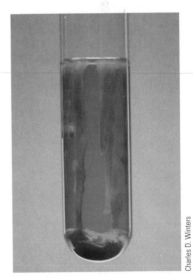

Precipitation of iron(III) hydroxide, $Fe(OH)_3$.

29 Metals in Qualitative Analysis

OUTLINE

A basic oxygen furnace. The addition of transition metals to molten steel converts it to stainless steels, which have many more uses.

Many of the metals whose ions are included in the qualitative analysis scheme for cations have been discussed previously in most general chemistry courses. For example, some important metals and their metallurgy were discussed in detail in Chapters 22 and 23 of *General Chemistry* by Whitten, Davis, Peck, and Stanley. Coordination compounds were covered in Chapter 25. The discussions that follow will refer to topics that are covered prior to this point in a typical first-year college chemistry course.

For each group of metals, we shall (1) tabulate some important properties, (2) discuss some of these properties, (3) indicate naturally occurring sources, (4) briefly describe their metallurgies, and (5) mention a few uses.

29-1 THE METALS OF ANALYTICAL GROUP I

Some properties of the metals of analytical Group I are tabulated in Table 29-1.

The analytical groups can be numbered in different ways, depending on which metals are included and which reactions are used in the analysis.

TABLE 29-1 *Some Properties of the Metals of Analytical Group I*

Metal	Atomic Number	Periodic Group Number	Atomic Weight	Atomic Radius (Å)	Important Oxidation States	Ionic Radius (Å)	Density at 20°C (g/cm³)	Compounds in Ores
copper	29	IB	63.546	1.28	$+1, +2$	Cu^{2+}, 0.70	8.9	Cu, Cu_2S, CuS, Cu_2O, CuO, $CuFeS_2$, $Cu(OH)_2 \cdot CuCO_3$
cadmium	48	IIB	112.411	1.54	$+2$	Cd^{2+}, 0.97	8.7	CdS
bismuth	83	VA	208.9804	1.50	$+3, +5$	Bi^{3+}, 0.96	9.8	Bi, Bi_2O_3, Bi_2S_3

Four pieces of jewelry made from copper minerals. Clockwise from top: malachite with azurite, malachite, azurite with chalcopyrite, and turquoise, $CuAl_6(PO_4)_4(OH)_8 \cdot 5H_2O$.

This bronze finial comes from Luristan (Persia) and dates from about 1200 BC.

The fact that copper utensils were used by Native Americans in the southern part of the United States is taken as evidence for trade among widely separated tribes.

Copper is used in the manufacture of U.S. coins, both pennies and cladded-copper dimes and quarters.

Cadmium sulfide, CdS, is used as a paint pigment.

Copper

Copper is a relatively soft, reddish yellow metal that is an excellent conductor of heat and electricity. Copper was known to many ancient civilizations, and it was probably the first metal used to fabricate tools and utensils.

Copper sometimes occurs as the free metal. The largest known deposits were near Houghton, Michigan. The largest piece found to date weighed more than 400 tons.

The commercially important ores of copper are copper(I) sulfide (Cu_2S, *chalcocite*) and a mixed sulfide ($CuFeS_2$, *chalcopyrite*), which contain less than 10% copper. Most known sources of the high-grade oxide ores—*cuprite*, Cu_2O, and *melaconite*, CuO—have been exhausted. The metallurgy of copper was discussed in Section 22-8 of *General Chemistry* by Whitten, Davis, Peck, and Stanley. Copper is refined electrolytically.

Copper is second in importance to iron, the most widely used metal. Copper has excellent thermal and electrical conductivity, chemical inertness, and usefulness in alloying elements. It is used extensively in alloys such as brass (Cu-Zn), bronze (Cu-Zn-Sn), sterling silver (Cu-Ag), aluminum bronze (Cu-Al), and German silver (Cu-Zn-Ni).

Copper pipes are used in plumbing because copper is easy to "work" and because it does not react with hot or cold water at an appreciable rate. Copper statues and roofs turn brown as a thin adherent film of copper oxide or copper sulfide forms. After long exposure to the atmosphere, they turn green due to the formation of basic copper(II) carbonate, $Cu(OH)_2 \cdot CuCO_3$.

Human need for trace amounts of copper has been demonstrated. Trace amounts of copper are found in plants that grow near deposits of copper ores. A copper compound (hemocyanin) serves the same oxygen-carrying function in the "blood" of lobsters and oysters that the iron compound hemoglobin serves in the blood of higher animals.

Cadmium

Cadmium is quite similar to zinc, but it is softer and more malleable and ductile. It is chemically less reactive than zinc.

$$Cd^{2+}(aq) + 2e^- \longrightarrow Cd(s) \qquad E^0 = -0.403 \text{ V}$$

It is used to plate iron when the need for a high-quality coating justifies the cost. Cadmium is used in nicad batteries and as a stabilizer in plastics. Cadmium sulfide is used as a paint pigment.

Cadmium is obtained primarily from zinc smelters and from the sludge produced by the electrolytic refining of zinc. Many zinc ores contain small amounts (<1%) of cadmium sulfide. A rare mineral, *greenockite*, contains a high percentage of CdS.

Cadmium is used in some low-melting alloys such as *Wood's metal* (mp = 65°C), which is 50% Bi, 25% Pb, 12.5% Sn, and 12.5% Cd. It is also used in the manufacture of some antifriction bearings (Cd-Ni) that have higher melting points than Babbit metal (Sb-Sn-Cu) bearings.

Cadmium-plated iron is more resistant to attack by salt water and alkaline solutions than is zinc-plated (galvanized) iron.

Bismuth

Bismuth is the heaviest of the known Group VA elements. Its physical and chemical properties are distinctly metallic. It is a hard, brittle metal with a reddish tint, and it exhibits the unusual property of expanding as it solidifies (as does antimony). It occurs as the free element as well as the oxide, Bi_2O_3, called *bismuth ocher*, and the sulfide, Bi_2S_3, called *bismuth glance*. Because bismuth occurs in many lead ores, its principal commercial source is as a by-product of the refining of lead.

The uses of bismuth are based on the facts that it expands when it solidifies and that some bismuth alloys have very low melting points. Where sharp, well-defined edges of castings are important (in type metals, for example), bismuth and antimony are used as alloying agents. Bismuth alloys are used in automatic sprinkler systems, electrical fuses, and safety plugs for boilers in which low-melting alloys are essential.

Electrical fuses are safety devices based on the low melting points of bismuth alloys.

29-2 THE METALS OF ANALYTICAL GROUP II

Some properties of the metals of analytical Group II are tabulated in Table 29-2.

Arsenic

Arsenic and antimony are Group VA metalloids. Arsenic is located just above, and antimony just below, the arbitrary line that separates metals and nonmetals. Arsenic exhibits

TABLE 29-2 *Some Properties of the Metals of Analytical Group II*

Metal	Atomic Number	Periodic Group Number	Atomic Weight	Atomic Radius (Å)	Important Oxidation States	Ionic Radius (Å)	Density at 20°C (g/cm³)	Compounds in Ores
arsenic	33	VA	74.9216	1.20	−3, +3, +5	As^{3+}, 0.58	5.7	As_2S_3, As_2S_2, FeAsS
tin	50	IVA	118.710	1.40	+2, +4	Sn^{2+}, 0.93 Sn^{4+}, 0.71	7.3	SnO_2
antimony	51	VA	121.75	1.40	+3, +5	Sb^{3+}, 0.76	6.7	Sb_2S_3

Arsenic sulfide ores: orpiment, As_2S_3; realgar, As_2S_2.

TABLE 29-3	Some Amphoteric Hydroxides	
Insoluble Amphoteric Hydroxide	**Complex Ion Formed in an Excess of a Strong Base**
$Be(OH)_2$	$[Be(OH)_4]^{2-}$
$Al(OH)_3$	$[Al(OH)_4]^-$
$Cr(OH)_3$	$[Cr(OH)_4]^-$
$Zn(OH)_2$	$[Zn(OH)_4]^{2-}$
$Sn(OH)_2$	$[Sn(OH)_3]^-$
$Sn(OH)_4$	$[Sn(OH)_6]^{2-}$
$Pb(OH)_2$	$[Pb(OH)_4]^{2-}$
$As(OH)_3$	$[As(OH)_4]^-$
$Sb(OH)_3$	$[Sb(OH)_4]^-$
$Si(OH)_4$	SiO_4^{4-} and SiO_3^{2-}
$Co(OH)_2$	$[Co(OH)_4]^{2-}$
$Cu(OH)_2$	$[Cu(OH)_4]^{2-}$

It has been established that animals, including humans, need small amounts of tin in their diets.

many properties that are characteristic of nonmetals, but it also exhibits some properties of metals. Arsenic(III) hydroxide (arsenous acid) is amphoteric (Table 29-3). Antimony is decidedly more metallic than arsenic, as is expected from its lower position in the periodic table.

Arsenic occurs in nature as the free element and in sulfide ores such as As_2S_3, *orpiment*; As_2S_2, *realgar*; and FeAsS, *arsenopyrite*. The sulfides are found in trace amounts in the sulfide ores of many metals. Most arsenic is obtained as a by-product of the production of other metals, notably copper.

The roasting of arsenic ores produces As_4O_6, which is reduced to elemental arsenic by heating with carbon.

$$As_4O_6(s) + 6C(s) \longrightarrow As_4(g) + 6CO(g)$$

The free arsenic is condensed and then purified by sublimation. Native arsenic ores are heated to sublime the arsenic, which is then purified by sublimation.

Most important uses of arsenic compounds are based on their poisonous nature. However, the need for small amounts of arsenic in the diets of animals has been established.

Antimony

Antimony is a brittle, lustrous metalloid with considerably more metallic character than arsenic. It is sometimes found as the free element. Most antimony is obtained from the sulfide ore, *stibnite*, in which Sb_2S_3 occurs as a black solid. The orange-red modification of this compound was used for cosmetic purposes at least 5000 years ago. Antimony exhibits amphoterism in both the +3 and +5 oxidation states. The metallurgy of antimony is similar to that of arsenic.

The principal uses of antimony depend on the facts that it expands when it solidifies (as does bismuth, with which it is used in type metal) and that it is relatively unreactive chemically. The "lead" plates used in lead storage batteries are 94% lead and 6% antimony. The small amount of antimony makes the plates much more resistant to attack by H_2SO_4.

Antimony alloys such as Babbit metal (Sb-Sn-Cu) are used extensively as antifriction machine bearings. Shrapnel is lead that has been hardened by the addition of 10 to 20% antimony.

Tin

Tin is a Group IVA element located near the arbitrary dividing line between metals and nonmetals. It is decidedly metallic in character, but its hydroxides are amphoteric (Table 29-3), which indicates some nonmetallic character.

Tin is relatively soft and ductile. Its excellent resistance to corrosion (reaction with H_2O and O_2) is the basis for its primary uses. Tin vessels have been found in ancient Egyptian tombs. The Romans and Phoenicians used tin that they obtained from *tinstone*, or *cassiterite*, SnO_2, deposits in England.

Tin is obtained commercially from cassiterite. The crushed ore is separated from the lighter rocky material and then roasted to remove arsenic and sulfur as volatile oxides. Oxides of other metals are extracted with hydrochloric acid, in which SnO_2 is insoluble. The remaining ore is then reduced with carbon in a furnace.

$$SnO_2(s) + 2C(s) \longrightarrow Sn(\ell) + 2CO(g)$$

Molten tin is drained off at the bottom of the furnace and cast into blocks. The blocks are then remelted so that the tin flows away from impurities that have higher melting points. Final purification is by electrolysis in a bath of sulfuric acid and hexafluorosilicic acid, $H_2[SiF_6]$. Pure tin cathodes and impure tin anodes are used. The process is similar to the electrolytic refining of copper.

Tin is used primarily in corrosion protection, that is, as tin plate for sheet iron. "Tin cans" are protected by a very thin layer of tin. Tin is also used in certain alloys such as bronze (Cu-Zn-Sn) and solder (Sn-Pb).

29-3 THE METALS OF ANALYTICAL GROUP III

Table 29-4 lists some properties of the metals of analytical Group III.

Aluminum is the only A group metal whose cations occur in analytical Group III. Therefore, the chemistry of the analytical Group III cations is largely the chemistry of d-transition metal cations. Most d-transition metal ions form coordination compounds readily.

The metals of analytical Group III are all classified as "active metals." The production of the free metals requires large amounts of energy. Refer to Appendix J for E^0 values.

Cobalt

Cobalt, like nickel and iron, is magnetic. It looks very much like iron except that it appears slightly pink. It is rendered passive by contact with concentrated nitric acid, as are iron and nickel.

Human need for trace amounts of cobalt also has been established.

There are deposits of cobalt ores containing CoAsS, *cobaltite*, and $CoAs_2$, *smaltite*. Most cobalt is obtained as a by-product of the metallurgy of copper, iron, nickel, silver, and other metals that occur as sulfide ores. After cobalt compounds have been converted to Co_3O_4 and isolated (quite a complex process), the oxide is reduced by metallic aluminum in a *highly exothermic process*.

$$3Co_3O_4(s) + 8Al(s) \xrightarrow{heat} 9Co(\ell) + 4Al_2O_3(s) + 4029 \text{ kJ/mol rxn}$$

The metal is purified electrolytically.

TABLE 29-4 *Some Properties of the Metals of Analytical Group III*

Metal	Atomic Number	Periodic Group Number	Atomic Weight	Atomic Radius (Å)	Important Oxidation States	Ionic Radius (Å)	Density at 20°C (g/cm³)	Compounds in Ores
aluminum	13	IIIA	26.98154	1.43	+3	Al^{3+}, 0.68	2.7	Al_2O_3
chromium	24	VIB	51.9961	1.27	+3, +6	Cr^{3+}, 0.61	7.1	$FeCr_2O_4$
manganese	25	VIIB	54.9380	1.26	+2, +3, +4, +7	Mn^{2+}, 0.80	7.2	MnO_2, Mn_2O_3, Mn_3O_4, mixed oxides
iron	26	VIIIB	55.847	1.26	+2, +3	Fe^{2+}, 0.75 Fe^{3+}, 0.64	7.9	Fe_2O_3, Fe_3O_4, $FeCO_3$, FeS
cobalt	27	VIIIB	58.9332	1.25	+2, +3	Co^{2+}, 0.72 Co^{3+}, 0.63	8.7	CoAsS, $CoAs_2$, CoS
nickel	28	VIIIB	58.69	1.24	+2, +4	Ni^{2+}, 0.70	8.9	NiS in mixed sulfides
zinc	30	IIB	65.39	1.38	+2	Zn^{2+}, 0.74	7.1	ZnS, ZnO, $ZnCO_3$

Crystals of nickel sulfate hexa-hydrate, $NiSO_4 \cdot 6H_2O$.

Cobalt is alloyed with iron and other metals to produce very hard materials that are used for high-speed cutting tools, surgical instruments, and other tools and instruments that must resist corrosion. Permanent magnets are made of alloys such as Alnico V (8% Al, 14% Ni, 24% Co, 3% Cu, and 51% Fe). Many cobalt compounds are used as contact catalysts.

Nickel

Nickel is a hard, malleable, ductile, silvery white, highly reflective metal that is resistant to corrosion. Like cobalt, it dissolves very slowly in dilute solutions of nonoxidizing acids such as HCl. It is readily soluble in $6\,M\,HNO_3$, but is rendered passive by concentrated HNO_3.

Metallic nickel and iron are believed to make up most of the core of the earth. Many meteorites are alloys of iron and nickel, which indicates that both metals are present in large quantities in celestial bodies.

Nickel is usually obtained from the ore *pentlandite*, a relatively abundant ore that contains a mixture of Ni, Cu, and Fe sulfides. The (Ni, Cu, Fe)S ore is roasted to convert the mixture of sulfides to oxides.

$$2(Ni, Cu, Fe)S + 3O_2 \longrightarrow 2(Ni, Cu, Fe)O + 2SO_2$$

The mixture of oxides is then reduced with carbon to produce a crude alloy known as **Monel metal.**

$$(Ni, Cu, Fe)O + C \longrightarrow (Ni, Cu, Fe) + CO$$

The composition of Monel metal is 60% Ni, 36% Cu, 3.5% Fe, and 0.5% Al. It is highly resistant to corrosion and is therefore widely used in industry.

Another preparation of nickel involves separation of the pulverized mixed sulfide ore by selective flotation. The NiS is then roasted and reduced with carbon as indicated in the previous reaction. This process yields metallic nickel that is about 95% pure. It can be purified electrolytically to give a product of better than 99.9% purity.

Alternatively, carbon monoxide can be passed over finely divided impure nickel at about 75°C to produce nickel carbonyl, $Ni(CO)_4$, a volatile (and very toxic) compound that boils at 43°C.

$$Ni(s) + 4CO(g) \xrightarrow{75°C} Ni(CO)_4(g)$$

The nickel carbonyl is condensed to a liquid and then decomposed by heat, in the reverse of the preceding reaction, to produce very pure nickel.

Because of its resistance to corrosion, nickel is used to plate iron, steel, and copper. Nickel plate offers the additional advantage of a highly reflective surface. *Nichrome* is an alloy (60% Ni, 25% Fe, and 15% Cr) that is used in the heating wires of electric heaters and toasters. There are many other examples of important alloys that contain nickel, such as the stainless steels used in cables, gears, and drive shafts.

Since 1974 we have known that trace amounts of nickel are essential for animal nutrition.

Iron

Iron is the second most abundant *metal* (4.7%) in the earth's crust after aluminum (7.5%). It is the most widely used of all metals. Iron was known as early as 4000 BC. It was used by primitive people because of its wide distribution and ease of reduction to the free metal. The red color of many rocks, clays, and soils is due to the presence of Fe_2O_3, the compound formed when iron rusts.

The two most abundant elements are *nonmetals*: oxygen (49.5%) and silicon (25.7%).

Iron is a silvery white metal of high tensile strength that takes a high polish well. It is ductile and soft compared with many other metals. The occurrence and metallurgy of iron were discussed in Section 22-7.

Iron is widely used because its properties can be modified by alloying with a variety of elements. Additionally, it can be protected from corrosion by many different kinds of coatings. For example, iron can be coated with (1) other metals such as Zn, Sn, Cu, Ni, Cr, Cd, and Pb; (2) ceramic materials such as those used in bathtubs and refrigerators; and (3) organic materials such as paint, lacquer, or asphalt.

Manganese

Manganese is a softer metal than iron. It is gray with a reddish tinge. Common ores of manganese are oxides, with *pyrolusite*, MnO_2, being the most important. Because manganese is used primarily to alloy with iron, the ore is usually reduced with coke in a blast furnace, which also reduces iron oxides. The impure mixture of metals is used to make steel.

$$MnO_2(s) + 2C(s) \longrightarrow Mn(impure) + 2CO(g)$$

Steel that contains 10 to 18% manganese is hard, tough, and resistant to wear. It is used to make rock-crushing machinery, armor plate, railroad rails, and similar items.

Since 1931 we have known that manganese is an essential trace element in the human diet.

Pyrite is iron(II) sulfide, FeS. Its brilliant gold color has earned it the nickname "fool's gold."

Chromium

Chromium is a lustrous, hard metal that is very resistant to corrosion. Like aluminum, it is a strong reducing agent,

$$Cr^{2+}(aq) + 2e^- \longrightarrow Cr(s) \qquad E^0 = -0.91 \text{ V}$$

that forms a thin, tough, protective film of oxide.

The most important ore of chromium is *chromite* (also called chrome iron ore), $FeCr_2O_4$. The ore is reduced by carbon in an electric furnace to form an alloy called *ferrochrome*.

$$FeCr_2O_4(s) + 4C(s) \longrightarrow \underbrace{Fe + 2Cr}_{ferrochrome} + 4CO(g)$$

Ferrochrome is used in making chromium steels. Stainless steel contains 14 to 18% Cr and is very resistant to corrosion. Chrome-vanadium steel contains 1 to 10% Cr together with about 0.15% vanadium. These very strong steels are used to make axles that must withstand constant strain as well as frequent shock and vibration.

Some chemistry of chromium was described in Section 23-10 of *General Chemistry* by Whitten, Davis, Peck, and Stanley. The need for trace amounts of chromium in the human diet was established in 1959.

Chromium is plated onto other metals to provide a shiny, protective coating.

Aluminum

Aluminum is the third most abundant element in the earth's crust (Table 1-3), and the most abundant metal. It is a Group IIIA element located just below the dividing line between metals and nonmetals. Not surprisingly, its hydroxide is amphoteric (Table 29-3).

Throughout this supplement, references to chapters or sections prior to Chapter 29 refer to *General Chemistry* by Whitten, Davis, Peck, and Stanley.

The metallurgy of aluminum was discussed in Section 22-6 of *General Chemistry* by Whitten, Davis, Peck, and Stanley. Although it is a very reactive metal,

$$Al^{3+}(aq) + 3e^- \longrightarrow Al(s) \qquad E^0 = -1.66 \text{ V}$$

aluminum quickly becomes coated with a thin, tough layer of oxide, Al_2O_3, that protects it from further atmospheric oxidation.

Aluminum is a relatively soft metal that can be extruded, that is, forced through a die by pressure and thereby formed into the desired shape. Frames for window screens and similar items are made from aluminum that has been shaped by extrusion. Alloys of aluminum that contain *d*-transition metals are much stronger and harder than aluminum itself.

Zinc

Zinc is a Group IIB element, and strictly speaking, it is not a *d*-transition metal. It is a reactive metal, but considerably less so than aluminum.

$$Zn^{2+}(aq) + 2e^- \longrightarrow Zn(s) \qquad E^0 = -0.76 \text{ V}$$

Like aluminum, zinc becomes coated with a protective oxide layer. This is then converted to blue-gray basic carbonate, $Zn(OH)_2 \cdot ZnCO_3$, on exposure to moist air.

Zinc hydroxide is amphoteric (Table 29-3). The metal dissolves both in nonoxidizing acids and in strong soluble bases with the evolution of hydrogen.

The main ore of zinc contains zinc sulfide, ZnS, and is called *zinc blende* or *sphalerite*. Cadmium, iron, lead, and arsenic sulfides usually occur in zinc sulfide ores.

Zinc sulfide ores are concentrated by flotation and then roasted.

$$2ZnS(s) + 3O_2(g) \longrightarrow 2ZnO(s) + 2SO_2(g)$$

Then the oxide is reduced by heating it with coal in a clay retort.

$$ZnO(s) + C(s) \longrightarrow Zn(\ell) + CO(g)$$

Zinc is sufficiently volatile (bp = 907°C) that it distills out of the retort as rapidly as it is produced. Impurities, primarily Cd, Fe, Pb, and As, are then separated by careful distillation of the crude zinc.

29-4 THE METALS OF ANALYTICAL GROUP IV

Some properties of the metals of analytical Group IV are tabulated in Table 29-5.

TABLE 29-5 *Some Properties of the Metals of Analytical Group IV*

Metal	Atomic Number	Periodic Group Number	Atomic Weight	Atomic Radius (Å)	Important Oxidation States	Ionic Radius (Å)	Density at 20°C (g/cm³)	Compounds in Ores
calcium	20	IIA	40.078	1.97	+2	Ca^{2+}, 1.14	1.6	$CaCO_3$, $CaCO_3 \cdot MgCO_3$, $CaSO_4$, CaF_2, mixed compounds
strontium	38	IIA	87.62	2.15	+2	Sr^{2+}, 1.32	2.6	$SrSO_4$, $SrCO_3$
barium	56	IIA	137.327	2.22	+2	Ba^{2+}, 1.49	3.5	$BaSO_4$, $BaCO_3$

(a)   (b)

Very pure calcium carbonate, $CaCO_3$, occurs in two crystalline forms, (a) calcite and (b) aragonite.

The cations that occur in analytical Group IV—Ca^{2+}, Sr^{2+}, and Ba^{2+}—are derived from three metals from Group IIA in the periodic table. These elements exhibit only the +2 oxidation state in their compounds. Sections 23-4, 23-5, and 23-6 contain information on the properties and occurrence, the reactions, and the uses of these metals and their compounds as well as those of magnesium, an analytical Group V cation.

Calcium

Calcium makes up about 3.4% of the earth's crust; it is the fifth most abundant element and the third most abundant metal. Large amounts of calcium occur in *limestone*, a rock that is primarily $CaCO_3$. *Dolomitic limestone* contains large amounts of $MgCO_3$ as well. Calcium carbonate occurs in other natural forms such as marine animal shells and pearls, as well as in deposits of *marl, marble, chalk, calcite* (often as very large crystals that are displayed in museums), and *aragonite*. Calcium sulfate occurs in deposits that contain *gypsum*, $CaSO_4 \cdot 2H_2O$, a hydrated compound, and also in the anhydrous compound *anhydrite*, $CaSO_4$. Calcium fluoride, CaF_2, occurs in a mineral called *fluorite* or *fluorspar*. Calcium occurs in many complex compounds such as *apatite*, which is usually formulated as $Ca_{10}(PO_4)_6X_2$ (where X = OH, Cl, or F).

Calcium, strontium, and barium are very strong reducing agents.

$$Ca^{2+}(aq) + 2e^- \longrightarrow Ca(s) \qquad E^0 = -2.87 \text{ V}$$
$$Sr^{2+}(aq) + 2e^- \longrightarrow Sr(s) \qquad E^0 = -2.89 \text{ V}$$
$$Ba^{2+}(aq) + 2e^- \longrightarrow Ba(s) \qquad E^0 = -2.90 \text{ V}$$

They are usually prepared by electrolysis of their molten chlorides.

$$MCl_2(\text{molten}) \xrightarrow{\text{electrolysis}} M(\ell) + Cl_2(g) \qquad (M = Ca, Sr, Ba)$$

Because Ca, Sr, and Ba are such powerful reducing agents, large amounts of energy are required to liberate the metals from their compounds.

Standard reduction potentials were determined for reactions that occur in aqueous solutions. Therefore, E^0 values are not strictly applicable to reactions in the absence of water.

Interestingly, calcium does not react with oxygen in room-temperature air at an appreciable rate. Strontium reacts with oxygen quite rapidly, and barium bursts into flames in moist air. Such kinetic differences could *not* be predicted from their E^0 values.

Strontium

Strontium is a rare element, and there are no commercial uses for the free element. Like calcium, it occurs in nature as the carbonate *strontianite*, $SrCO_3$, and as the sulfate *celestite*, $SrSO_4$. Both are quite rare and somewhat less soluble than the corresponding calcium compounds.

Barium

Barium also occurs in nature as the carbonate *witherite*, $BaCO_3$, and as the sulfate *barite*, or *heavy spar*, $BaSO_4$. The solubility of $CaCO_3$ is intermediate between the solubilities of $BaCO_3$ and $SrCO_3$, whereas $BaSO_4$ is much less soluble than $SrSO_4$, which in turn is less soluble than $CaSO_4$.

Barium is produced both by electrolysis of its molten chloride and by reduction of a mixture of barium oxide and barium peroxide with aluminum in a vacuum furnace at 1000 to 1100°C.

$$\left.\begin{array}{l} BaO \\ BaO_2 \end{array}\right\} \xrightarrow{\text{Al}} Ba(g) + Al_2O_3(s)$$

Barium is a stronger reducing agent than aluminum, and we might expect that this reaction would not be possible. However, LeChatelier's Principle comes to the rescue again. The furnace is operated at very low pressures (10^{-3} to 10^{-4} torr) so that barium vaporizes and distills out of the furnace as rapidly as it is formed. Aluminum oxide is thermodynamically a very stable compound ($\Delta H_f^0 = -1676$ kJ/mol, $\Delta G_f^0 = -1562$ kJ/mol). The high thermodynamic stability of Al_2O_3 and the volatility of Ba at 1000 to 1100°C make the process viable.

29-5   THE METALS OF ANALYTICAL GROUP V

Some properties of the metals of analytical Group V are tabulated in Table 29-6. Magnesium, Mg^{2+}; sodium, Na^+; potassium, K^+; and ammonium, NH_4^+, ions, the so-called soluble group, constitute analytical Group V. Ammonium ion is a common cation that is not

TABLE 29-6 *Some Properties of the Metals of Analytical Group V*

Metal	Atomic Number	Periodic Group Number	Atomic Weight	Atomic Radius (Å)	Important Oxidation States	Ionic Radius (Å)	Density at 20°C (g/cm³)	Compounds in Ores
magnesium	12	IIA	24.3050	1.60	+2	Mg^{2+}, 0.85	1.7	$MgCl_2$ (brines), $MgCl_2$, $MgSO_4$ (sea water)
sodium	11	IA	22.98977	1.86	+1	Na^+, 1.16	0.97	NaCl, complex compounds
potassium	19	IA	39.0983	2.27	+1	K^+, 1.52	0.86	KCl, complex compounds

derived from a metal; the group separations place it in analytical Group V because the solubilities of its compounds are similar to those of the potassium ion. Magnesium is a Group IIA metal; sodium and potassium are found in Group IA.

Sections 23-4, 23-5, and 23-6 in *General Chemistry* by Whitten, Davis, Peck, and Stanley contain information on the alkaline earth metals. Section 22-5 describes the commercial recovery of magnesium from sea water and some of its important uses.

Magnesium

Magnesium is the eighth most abundant element in the earth's crust (1.9%) and is less abundant than sodium (2.6%) and potassium (2.4%). Because both $MgCl_2$ and $MgSO_4$ are soluble in water, magnesium ions frequently occur in ground waters as well as in oceans and lakes.

In recent years, the problem of magnesium deficiency in the diets of lactating cows has come to light, and some progress has been made in dealing with it. The effects are apparently quite complex and so far are only partially understood. What is known follows. Cows that have young suckling calves and that graze on lush new pasture in the early spring sometimes develop a disorder called *grass tetany*. It is usually fatal unless detected and treated within a few hours. Research has shown that the disease usually develops in apparently healthy animals that have insufficient magnesium in their diets. Two methods have been partially successful in combating this phantom killer of lactating cows. The application of dolomitic limestone ($CaCO_3 \cdot MgCO_3$) to pasturelands significantly reduces the incidence of grass tetany, as does the inclusion of magnesium compounds in the minerals and dry feeds that are fed to cattle.

The costs of beef and dairy products depend on the effectiveness with which agriculture scientists combat problems such as grass tetany.

Sodium and Potassium

Sodium and potassium are widely distributed in nature, being the sixth and seventh most abundant elements (2.6% and 2.4%), respectively, in the earth's crust. The properties, occurrence, and reactions of the alkali metals were discussed in Sections 23-1, 23-2, and 23-3 of *General Chemistry* by Whitten, Davis, Peck, and Stanley.

Both sodium and potassium can be prepared by electrolysis of anhydrous molten compounds such as NaCl, NaOH, KCl, and KOH. In the production of sodium by the electrolysis of molten NaCl, some $CaCl_2$ is added to the electrolysis mixture to lower its melting point from about 800°C to about 600°C. This significantly reduces the amount of energy required to produce sodium.

Potassium is more expensive than sodium. Because sodium can be used very effectively in most industrial processes, very little potassium is produced today. As an alternative to electrolysis of molten KCl or KOH, potassium is produced by the reaction between molten sodium and molten KCl.

$$Na(\ell) + KCl(\ell) \xrightarrow{\text{heat}} K(g) + NaCl(\ell)$$

This is another example of a reaction that we might not expect to occur, but that is used in an industrial process. Recall that E^0 values are strictly applicable only to reactions in aqueous solutions. Apparently, the greater volatility of potassium compared with sodium is an important factor (another application of LeChatelier's Principle) in this reaction.

Exercises

Analytical Group I Metals

1. (a) Write the ground state electron configuration ($\uparrow\downarrow$ notation and shorthand $ns^x np^y nd^z$ notation) for each metal in analytical Group I. (b) Write the ground state electron configuration for each Group I cation.
2. List the names of the important ores of the analytical Group I metals and the formulas for the commercially important compound(s) in each ore.
3. Many important ores of the analytical Group I metals are sulfide ores. What does this suggest about the stability of these compounds?
4. Briefly describe the metallurgy of each metal in analytical Group I. Write equations for the important reaction(s).
5. List some important uses for each metal (or its compounds) in analytical Group I.
6. The largest mass of native copper discovered to date is 420 short tons. It was found in northern Michigan. The density of copper is 8.65 g/cm^3 at 20°C. (a) If this mass were a perfect cube, how long would each edge be? (b) If it were a perfect sphere, what would its diameter be?

Analytical Group II Metals

7. (a) Write the electron configuration ($\uparrow\downarrow$ notation and shorthand $ns^x np^y nd^z$ notation) for each metal in analytical Group II. (b) Write the electron configuration for each Group II cation.
8. List the names of the important ores of the analytical Group II metals and the formulas for the commercially important compound(s) in each ore.
9. The important ores of the analytical Group II metals are sulfide or oxide ores. What does this suggest about the stability of these compounds?
10. Briefly describe the metallurgy of each metal in analytical Group II. Write equations for the important reaction(s).
11. List some important uses for each metal (or its compounds) in analytical Group II.

Analytical Group III Metals

12. (a) Write the electron configuration ($\uparrow\downarrow$ notation and shorthand $ns^x np^y nd^z$ notation) for each metal in analytical Group III. (b) Write the electron configuration for each Group III cation.
13. List the names of the important ores of the analytical Group III metals and the formulas for the commercially important compound(s) in each ore.

14. The important ores of the analytical Group III metals are sulfide or oxide ores. What does this suggest about the stability of these compounds?
15. Briefly describe the metallurgy of each metal in analytical Group III. Write equations for the important reaction(s).
16. List some important uses for each metal (or its compounds) in analytical Group III.

Analytical Group IV Metals

17. (a) Write the electron configuration ($\uparrow\downarrow$ notation and shorthand $ns^x np^y nd^z$ notation) for each metal in analytical Group IV. (b) Write the electron configuration for each Group IV cation.
18. List the names of the important ores of the analytical Group IV metals and the formulas for the commercially important compound(s) in each ore.
19. The important ores of the analytical Group IV metals are carbonate and sulfate ores. What does this suggest about the stability of these compounds?
20. (a) List the carbonates of the analytical Group IV cations in order of increasing solubility in water. (b) List the sulfates of the analytical Group IV cations in order of increasing solubility in water.
21. Briefly describe the metallurgy of each metal in analytical Group IV. Write equations for the important reaction(s).
22. List some important uses for each metal (or its compounds) in analytical Group IV.
23. (a) How is it possible that aluminum, a less active metal, can displace barium, a more active metal, from its molten compounds? (b) If metallic aluminum were placed in an aqueous $BaCl_2$ solution, would you expect metallic barium to be formed? Why?

Analytical Group V Metals

24. (a) Write the electron configuration ($\uparrow\downarrow$ notation and shorthand $ns^x np^y nd^z$ notation) for each metal in analytical Group V. (b) Write the electron configuration for each Group V metal cation.
25. List the names of the important ores of the analytical Group V metals and the formulas for the commercially important compound(s) in each ore.
26. (a) How can large concentrations of NaCl and KCl exist in sea water and brines? (b) Would you expect to find high concentrations of AgCl in sea water or brines? Why?

Introduction to Laboratory Work

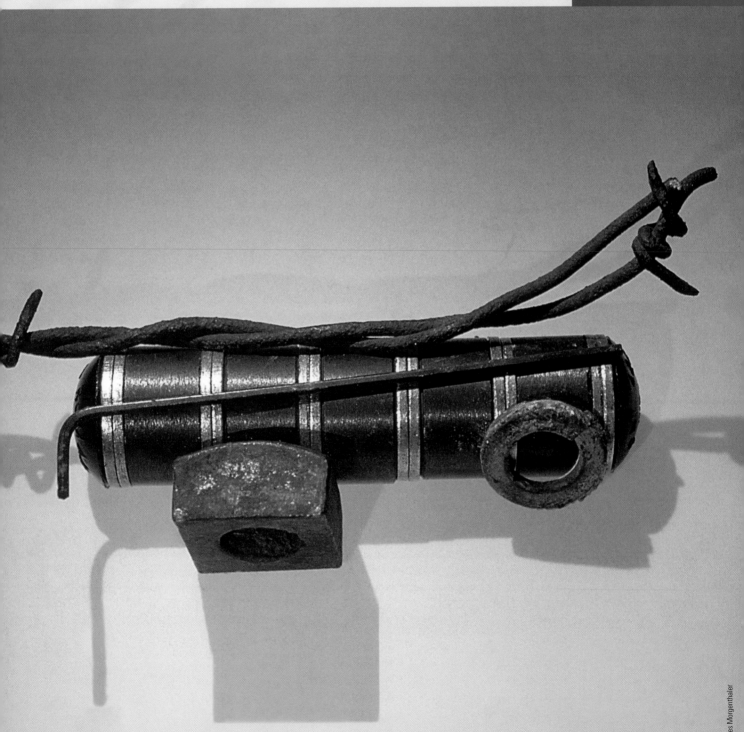

James Morgenthaler

The "cow magnet" shown here is made of Alnico, an alloy of Al, Ni, and Co. As the cow grazes, she picks up "scrap iron," which is attracted to the magnet in her first stomach. This prevents "scrap iron" from being carried into the cow's digestive tract, where it could cause serious damage.

OUTLINE

I n qualitative analysis we learn to make observations accurately and to interpret them logically. These may be the most important skills we develop in the educational process. We shall see the effects of various kinds of equilibria as we observe chemical reactions. We shall observe (1) the precipitation of insoluble compounds and (2) the dissolution of insoluble compounds by complex ion formation, by formation of weak electrolytes, and by oxidation–reduction reactions. We shall also learn careful laboratory manipulations.

30-1 BASIC IDEAS OF QUALITATIVE ANALYSIS

Qualitative analysis is concerned with identification of the ions that are present in a substance. Initially we work with solutions that contain only limited numbers of ions in specific groups. Then we work with more complex mixtures. We separate ions with similar properties into small groups.

We separate the members of the small groups so that we can isolate and identify each one. *Group separations* are based on similar chemical properties, whereas most *identifications* of individual ions are based on properties that are different. The common cations are listed in Figure 30-1, which shows how they can be separated into groups. The order of these operations has been chosen so that the substances added in each step will not interfere in the later steps.

In analyzing *general unknowns* we precipitate analytical Groups I and II together. Then we dissolve the analytical Group II sulfides in a strongly basic solution (4 *M* NaOH). The analytical Group I sulfides are insoluble in this solution. A *known* solution contains all the cations in a group. An *unknown* solution may contain any or all of the cations in the group.

The analytical groups can be numbered in different ways, depending on which metals are included and which reactions are used in the analysis. It is important to recognize that an analytical group number does *not* refer to a group of elements in the periodic table.

1150

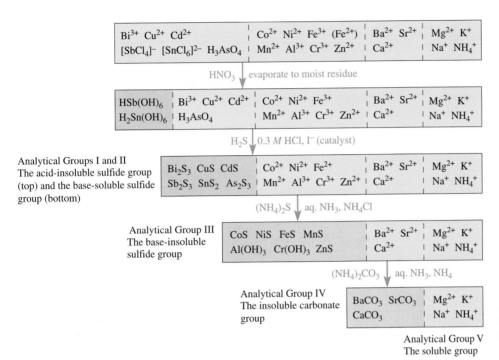

Analytical Groups I and II
The acid-insoluble sulfide group
(top) and the base-soluble sulfide
group (bottom)

Analytical Group III
The base-insoluble
sulfide group

Analytical Group IV
The insoluble carbonate
group

Analytical Group V
The soluble group

Figure 30-1 Flow chart for the systematic separation of common cations into groups. Blue areas represent precipitated species.

30-2 CATIONS OF ANALYTICAL GROUP I

To analyze a solution that contains analytical Group I cations (Bi^{3+}, Cu^{2+}, and Cd^{2+}), we start with a solution that contains either chloride or nitrate salts of these cations. It is made 0.3 M in HCl and saturated with hydrogen sulfide. Under these conditions the analytical Group I cations precipitate as insoluble sulfides. They are bismuth(III), copper(II), and cadmium sulfides. These ions are called the **acid-insoluble sulfide group** to distinguish them from the Group II cations.

30-3 CATIONS OF ANALYTICAL GROUP II

The conditions under which the analytical Group II sulfides are precipitated are similar to those for analytical Group I. (*In general unknowns, analytical Group I and Group II cations will be precipitated together.*) However, antimony and arsenic are both in the +5 oxidation state after the solution has been evaporated with HNO_3. The sulfides As_2S_3 and Sb_2S_3 are easier to precipitate and work with than are As_2S_5 and Sb_2S_5. We add a few drops of NH_4I to reduce both arsenic and antimony to the +3 oxidation state.

An important difference between Group I sulfides and Group II sulfides is the solubility of the Group II sulfides in excess 4 M NaOH solution. The Group I sulfides are insoluble in this solution.

30-4 CATIONS OF ANALYTICAL GROUP III

The term *basic-insoluble sulfide group* is not quite accurate because five sulfides and two hydroxides are precipitated in analytical Group III. However, the term is traditional.

After the analytical Groups I and II sulfides have been removed, the solution that contains the remaining cations is again saturated with hydrogen sulfide. Then an excess of aqueous ammonia is added. The sulfides of cobalt(II), nickel(II), manganese(II), iron(II), and zinc, together with the hydroxides of aluminum and chromium(III), precipitate from the buffered basic solution that contains ammonium sulfide. These ions are called the **basic-insoluble sulfide group.**

30-5 CATIONS OF ANALYTICAL GROUP IV

The next group of cations—barium, strontium, and calcium ions—includes three metals in the same family (group) in the periodic table. They are precipitated as carbonates by the addition of ammonium carbonate to the buffered basic solution from which the analytical Group III ions have been removed. This group is known as the **insoluble carbonate group.**

30-6 CATIONS OF ANALYTICAL GROUP V

The solution from which the analytical Group IV cations have been removed now contains only the **soluble group** cations. These are sodium, potassium, magnesium, and ammonium ions. They are known as the soluble group because they cannot be precipitated by a single reagent.

30-7 COMMENTS ON LABORATORY MANIPULATIONS

Reagent refers to any substance that is used to analyze or identify another substance. Many are kept in bottles on the laboratory shelf; others must be prepared just before use or even within the solution to be analyzed (in situ).

In **semimicro** qualitative analysis, we use small volumes in small test tubes so that we can separate solids from liquids rapidly with a centrifuge. Small volumes of liquids are conveniently removed from centrifuged solids with capillary pipets. Liquid reagents are added with medicine droppers or dropping bottles.

Your instructor will supply a list of laboratory apparatus. When a desk has been assigned, check the apparatus in the desk according to the instructions given by your instructor. Obtain any necessary replacements from the stockroom. Wash all of the equipment thoroughly with detergent solution, rinse it thoroughly with tap water, and finally rinse it with distilled water. All equipment must be *clean* because traces of impurities interfere with many qualitative tests.

30-8 CAPILLARY PIPETS

Liquid reagents may be added with standard medicine droppers or directly from the dropping bottles on the shelves in the laboratory. Ordinary medicine droppers deliver about 1 mL per 20 drops. Capillary pipets, with smaller tips, deliver approximately 1 mL per 40 drops.

Capillary pipets are used to transfer liquids from one test tube to another. If capillary pipets are not available, prepare several from glass tubing. (Your instructor may demonstrate such glass-working operations as cutting and fire-polishing.) Choose a 15-cm to

18-cm length of 8-mm-diameter soft glass tubing, and heat it over the Bunsen flame with rotation until the middle section of the glass softens. Remove the tube from the flame, hold it vertically, and slowly draw it out until the bore (the center opening) is approximately 1 mm across (Figure 30-2).

After the tube has cooled, cut the capillary at the midpoint and fire-polish the capillary ends. Flare the wide ends of the tubes by heating them until they are soft and then quickly pressing them down against a flat metal surface. After the pipets are cool, attach medicine-dropper bulbs to the flared ends.

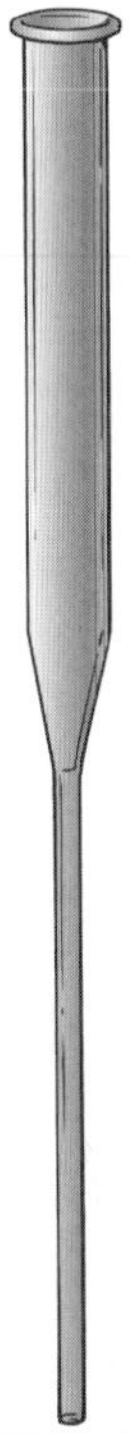

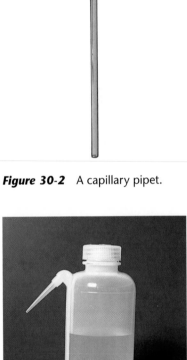

Figure 30-2 A capillary pipet.

> **C A U T I O N !**
>
> *Hot glass looks exactly like cold glass!* Even experienced lab technicians sometimes burn themselves by picking up hot glassware too soon. If in doubt, use a towel or tongs to pick up glass.

30-9 STIRRING RODS

Use a 3-mm soft glass rod (not tubing) to make at least six stirring rods, approximately 12 to 15 cm in length. Fire-polish both ends of each rod.

Only distilled water should be used in the analytical procedures because tap water contains ions such as Ca^{2+}, Mg^{2+}, Fe^{3+}, Fe^{2+}, HCO_3^-, SO_4^{2-}, and Cl^-. These ions are among those tested for in the unknown solutions. Several of them interfere with tests for other ions. Keep a wash bottle (Figure 30-3) filled with distilled water.

30-10 REAGENTS

The small dropping bottles in your desk are used to store commonly used reagents. Remove the droppers, remove the bulbs from the droppers, and wash the bottles and droppers thoroughly in hot detergent solution. Rinse them well with tap water and finally with distilled water. Reassemble the droppers and place them in the bottles. Label the bottles as follows:

1. $1.0\,M\,NH_3$ **4.** $0.1\,M\,HCl$ **7.** $6.0\,M\,H_2SO_4$

2. $3.0\,M\,NH_3$ **5.** $6.0\,M\,HCl$ **8.** $4.0\,M\,NaOH$

3. $6.0\,M\,NH_3$ **6.** $6.0\,M\,HNO_3$ **9.** 5% thioacetamide

Your instructor will show you where bottles filled with these reagents are located. Use them to fill your dropping bottles. The tenth dropping bottle is filled with distilled water. Label it. You will also be shown other reagents that you will use less frequently. There is another group of bottles filled with *known solutions* on the reagent shelf. Note that each ion is available in an individual bottle, and there are *group knowns* as well.

Figure 30-3 A wash bottle should be filled with distilled water. Use it to rinse your glassware.

30-11 PRECIPITATION

We use precipitate formation as the usual method for group separations. Precipitations are carried out in small test tubes. It is imperative that all separations be *as complete as possible.*

When a precipitation reaction is carried out, one must always check for *completeness of precipitation*. This is done by centrifuging the reaction mixture at high speed for approximately 1 minute (Section 30-12), adding one drop of the precipitating reagent to the clear solution, and observing carefully to see whether additional precipitate forms (Figure 30-4). If the addition of one more drop of the precipitating reagent shows that precipitation is incomplete, the mixture should be stirred thoroughly, centrifuged for approximately 1 minute, and tested for complete precipitation a second time.

Ions from an earlier group that remain in solution may interfere with the analysis of a later group.

Collision of ions, atoms, or molecules is a necessary condition for reaction.

Because precipitation reactions are carried out in small test tubes, considerable effort is necessary to ensure that reactants are mixed intimately. One effective way is to place a stirring rod inside the test tube, hold the top of the test tube between the thumb and index finger, and shake it vigorously sideways so that it strikes the little finger.

This motion imparts rotary motion to the liquid inside the test tube, and the presence of the stirring rod assists in intimate mixing of the contents.

Many precipitation reactions result in the formation of **colloidal particles,** particles that are too small to be separated from the liquid phase by filtration or by centrifuging the mixture. Colloidal suspension can usually be coagulated by heating the mixture in a water bath for a few minutes. This process is called **digestion.** As the mixture is warmed, the very small particles dissolve and recrystallize on the surfaces of larger particles. The presence of a colloidal suspension is detected from the opaqueness of the solution; that is, the solution is not transparent. When a colloidal suspension has coagulated, the solution (the **supernatant liquid**) becomes transparent, or clear (Figure 30-5). "Clear" and "colorless" do not have the same meaning. A *clear* solution is transparent; a *colorless* solution has no color. All true solutions are clear; many are colored.

After a colloidal suspension has coagulated, the mixture is centrifuged and the liquid is withdrawn with a capillary pipet. A slight excess of precipitating reagent is always added to reduce the solubility of the precipitate by the common ion effect. However, a very large excess of precipitating reagent should be avoided because it may form soluble compounds containing complex ions.

James Morgenthaler

Figure 30-4 Testing for complete precipitation. The solution in the centrifuged mixture should be clear (*left*). Add one drop of the precipitating reagent. The formation of more precipitate shows that precipitation was incomplete (*right*). Repeat the procedure until no more precipitate forms.

James Morgenthaler

Figure 30-5 A colloidal suspension appears cloudy or opaque (*left*). A true solution with a precipitate at the bottom of the test tube (*right*).

30-12 CENTRIFUGATION OF PRECIPITATES AND TRANSFER OF THE LIQUID (CENTRIFUGATE)

Precipitates are separated from solutions by centrifugation (Figure 30-6). The mixture is rotated at a very high speed so that the denser precipitate is forced to the bottom of the test tube. A balance tube containing a volume of water equal to the volume of solution to be centrifuged is always placed in the centrifuge exactly opposite the solution of interest. A gummed label should be placed at the top of the balance tube so that it can always be identified easily.

After a mixture has been centrifuged, the tube should be held at an angle so that the liquid (the **centrifugate**) can be easily drawn into a capillary pipet (Figure 30-7). Avoid disturbing the precipitate as the liquid is withdrawn. On occasion, bits of precipitate may be drawn into the capillary pipet with the liquid. To prevent this, wind a very small piece of cotton around the end of the capillary pipet to serve as a filter.

An unbalanced load in the centrifuge head will cause severe vibration and damage the centrifuge's bearings.

Supernate is often used to describe the clear solution above a precipitate. However, the term *centrifugate* is more descriptive in this case.

30-13 WASHING PRECIPITATES

A precipitate is always wet with the liquid from which it was separated. Precipitates are washed with small amounts of liquid to remove as much of the adhering liquid as possible. Use a stirring rod to break up the precipitate after the wash liquid is added, stir thoroughly, and centrifuge the mixture. When the solution that was initially separated from the precipitate is to be used in subsequent steps, the first wash liquid should be added to that solution. Usually a precipitate is washed twice, and the second wash liquid is discarded.

30-14 DISSOLUTION AND EXTRACTION OF PRECIPITATES

Figure 30-6 A centrifuge.

On occasion, it is desirable to dissolve only a part of a precipitate; that is, certain components of a solid mixture are dissolved but others are not. This process is called **extraction.** It is necessary to break up the precipitate with a stirring rod so that the dissolving liquid comes into intimate contact with the precipitate. Often the mixture is heated, because most substances are more soluble at higher temperatures and most reactions occur more rapidly at higher temperatures. The mixture must always be stirred thoroughly and frequently during extraction. It is then centrifuged and separated.

30-15 HEATING MIXTURES AND SOLUTIONS

Often it is necessary to heat a mixture to cause precipitation or to dissolve or extract precipitates. Keep a water bath gently boiling throughout the laboratory period. When it is necessary to heat a solution, the test tube can be placed into the hot water bath immediately.

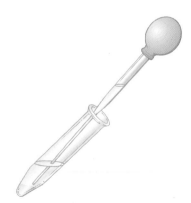

Figure 30-7 A capillary pipet is used to separate the liquid (centrifugate) from the precipitate.

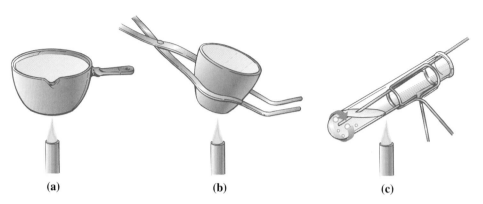

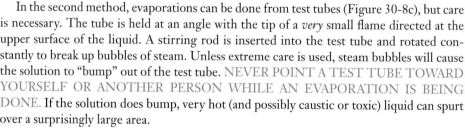

Figure 30-8 Evaporation of small amounts of liquid from (a) a casserole, (b) a crucible, and (c) a test tube.

30-16 EVAPORATION

Often we wish to reduce the volume of a solution by evaporating some of the liquid, or to evaporate a solution to a moist residue or even to dryness. This may be accomplished in two ways. In the first method the solution is placed in a small evaporating dish or casserole and heated gently until the liquid has evaporated (Figure 30-8a, b). Extreme caution must be exercised in evaporating small amounts of liquid because very little heat is required. Once an evaporating dish or casserole becomes hot, it may remain above the boiling point of water for some time after the flame has been removed. In most cases it is undesirable to bake residues.

In the second method, evaporations can be done from test tubes (Figure 30-8c), but care is necessary. The tube is held at an angle with the tip of a *very* small flame directed at the upper surface of the liquid. A stirring rod is inserted into the test tube and rotated constantly to break up bubbles of steam. Unless extreme care is used, steam bubbles will cause the solution to "bump" out of the test tube. NEVER POINT A TEST TUBE TOWARD YOURSELF OR ANOTHER PERSON WHILE AN EVAPORATION IS BEING DONE. If the solution does bump, very hot (and possibly caustic or toxic) liquid can spurt over a surprisingly large area.

> ⚠ When the solution in a test tube "bumps," it sometimes surprises a student and causes the student to drop the test tube.

30-17 KNOWN SOLUTIONS

Remember that known solutions contain all the cations in the group.

Known solutions should be analyzed before unknown solutions are analyzed. This enables you to learn the details of the analytical procedure and to recognize the expected results of tests for ions. The textures of precipitates vary widely. Crystalline precipitates are usually much denser than amorphous precipitates, so there appears to be relatively little crystalline precipitate and much more amorphous solid when equimolar quantities are present. Such observations are important. Make careful notes as known solutions are analyzed. A "known report" or an "unknown report" should be filled out as each solution is analyzed.

Colors are distinctive, and one must not rely on verbal descriptions of colors. Different people see colors somewhat differently, and therefore it is important to *know* how each colored substance appears *to you*.

The concentrations of cations in both known and unknown solutions are approximately 5 mg cation/mL of solution.

30-18 UNKNOWN SOLUTIONS

After the analysis of a known solution has been completed satisfactorily and the results have been approved by your instructor, ask the instructor to give you an unknown solution. The instructor may wish to ask several questions at this point to make sure you understand the analytical procedures.

The unknown may contain any or all of the ions in the group or groups. There are no unknowns that contain no ions, and no dyes have been added. The color of an unknown may give valuable clues about the presence or absence of certain ions. This is an important, legitimate observation. The results obtained for an unknown must always be consistent with the color of the unknown. Keep in mind that a mixture of colored ions may have a color that is not characteristic of any of the individual ions.

30-19 UNKNOWN REPORTS

Your instructor will indicate the kind of notebook you should use to record the results of the analyses. Observations should be recorded in your notebook *in ink* as soon as they are made. Equations should be written to describe the behavior of each ion as the various tests are run. Your instructor will specify the kind of report form to be used in reporting the analysis of unknowns.

30-20 LABORATORY ASSIGNMENTS

A list of suggested laboratory assignments is given here. Your instructor will indicate the number of unknowns required and the dates they are due.

1. Construct capillary pipets, stirring rods, and a wash bottle if necessary. Fill individual reagent bottles.

Analyze solutions containing the following ions:

Known	Unknown
2. The cations of Group I	**3.** Some Group I cations
4. The cations of Group II	**5.** Some Group II cations
6. The cations of Groups I and II	**7.** Some Groups I and II cations
8. The cations of Group III	**9.** Some Group III cations
10. The cations of Group IV	**11.** Some Group IV cations
12. The cations of Group V	**13.** Some Group V cations
	14. Some cations of Groups I to V

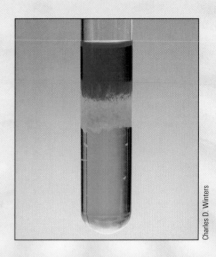

Charles D. Winters

31-1 COMMON OXIDATION STATES OF METALS IN CATION GROUP I

Nearly all compounds of bismuth that are stable in aqueous solution contain bismuth in the +3 oxidation state. A few compounds of bismuth(V) are known, but Bi(V) is a very powerful oxidizing agent in acidic solutions.

Copper exists in two oxidation states in aqueous solutions: Cu(I) and Cu(II). The following standard reduction potentials show the relative stabilities of simple copper(I) and copper(II) compounds in aqueous solutions.

$$Cu^+(aq) + e^- \longrightarrow Cu(s) \qquad E^0 = +0.521 \text{ V}$$
$$Cu^{2+}(aq) + 2e^- \longrightarrow Cu(s) \qquad E^0 = +0.337 \text{ V}$$
$$Cu^{2+}(aq) + e^- \longrightarrow Cu^+(aq) \qquad E^0 = +0.153 \text{ V}$$

Combination of the appropriate half-reactions gives

$$2Cu^+(aq) \longrightarrow Cu^{2+}(aq) + Cu(s) \qquad E^0_{cell} = +0.368 \text{ V}$$

This tells us that simple compounds containing copper(I) are unstable with respect to disproportionation into Cu(II) and metallic copper.

Cadmium exhibits only the +2 oxidation state in aqueous solutions.

One test for copper(II) ions involves the reaction with excess aqueous ammonia. To minimize mixing, concentrated aqueous NH_3 was added slowly to a solution of copper(II) sulfate, $CuSO_4$. Unreacted blue copper(II) sulfate solution remains in the bottom part of the test tube. The light blue precipitate in the middle is copper(II) hydroxide, $Cu(OH)_2$. The top layer contains deep blue $[Cu(NH_3)_4](OH)_2$, which was formed as some $Cu(OH)_2$ dissolved in excess aqueous NH_3.

However, as we shall see, Cu(I) can be stabilized in anionic complexes such as $[Cu(CN)_2]^-$.

31-2 INTRODUCTION TO THE ANALYTICAL PROCEDURES

The ions of analytical Group I are referred to as the **acid-insoluble sulfide group** because they are precipitated as sulfides from a saturated solution of H_2S that is also 0.30 M in HCl. These sulfides are insoluble in sodium sulfide solution. Bismuth(III), copper(II), and cadmium ions form sulfides that are insoluble in such solutions.

Thioacetamide serves as a source of hydrogen sulfide. It hydrolyzes in hot acidic solutions to produce hydrogen sulfide and ammonium acetate.

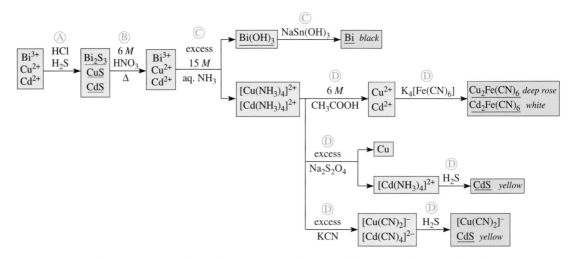

Figure 31-1 Analytical Group I flow chart. Circled letters refer to analytical steps. Species formed in confirmatory tests are shown with blue backgrounds. Precipitated species are underlined.

$$CH_3-\overset{\overset{\displaystyle S}{\|}}{C}-NH_2 + 2H_2O \xrightarrow{\text{heat}} \underbrace{CH_3\overset{\overset{\displaystyle O}{\|}}{C}-O^- + NH_4^+}_{\substack{\text{usually written as} \\ NH_4CH_3COO}} + H_2S$$

thioacetamide

When directions call for the addition of thioacetamide followed by heating, this amounts to the addition of hydrogen sulfide to the solution. The sulfides of analytical Group I are quite insoluble, as Table 31-1 indicates.

The flow chart in Figure 31-1 outlines schematically the precipitation of the analytical Group I sulfides and the analysis of this group of cations.

TABLE 31-1	*Solubility Products for Group I Sulfides*

Sulfide	K_{sp}
Bi_2S_3	1.6×10^{-72}
CuS	8.7×10^{-36}
CdS	3.6×10^{-29}

A more extensive table of solubility products appears in Appendix H. Throughout this supplement, references to chapters or sections prior to Chapter 29 refer to *General Chemistry* by Whitten, Davis, Peck, and Stanley.

31-3 PREPARATION OF SOLUTIONS OF ANALYTICAL GROUP I CATIONS

Solutions of the analytical Group I ions may be prepared by dissolving nitrates in dilute nitric acid to prevent hydrolysis of the metal ions (Section 18-11 of *General Chemistry* by Whitten, Davis, Peck, and Stanley). Solutions of copper(II) nitrate are blue. Solutions of cadmium and bismuth(III) nitrate are colorless.

Solutions of Cu^{2+}, Cd^{2+}, and Bi^{3+} ions may also be prepared by dissolving their chlorides in dilute hydrochloric acid. Solutions of copper(II) chloride are green; the others are colorless.

31-4 SEPARATION OF ANALYTICAL GROUPS I AND II CATIONS FROM GROUP III AND SUBSEQUENT GROUP CATIONS

Cadmium sulfide is the most soluble of the analytical Group I sulfides, $K_{sp} = 3.6 \times 10^{-29}$ (see Table 31-1). Zinc sulfide is the least soluble of the freshly precipitated Group III sulfides, $K_{sp} = 1.1 \times 10^{-21}$ (Table 33-1). Therefore, to effect a clean separation of the Groups I and II from Group III cations, the acidity must be adjusted so that cadmium sulfide is precipitated as completely as possible in Group I and the solubility product for zinc sulfide is not exceeded. The following calculations illustrate this point.

EXAMPLE 31-1   *Equilibria in Sulfide Precipitations*

What is the minimum $[H^+]$ that will prevent the precipitation of zinc sulfide in a solution that is 0.020 M in $Zn(NO_3)_2$ (approximately 1.3 mg Zn^{2+}/mL) and saturated with H_2S?

Plan

In Appendix H we find K_{sp} for ZnS.

$$K_{sp} = [Zn^{2+}][S^{2-}] = 1.1 \times 10^{-21}$$

We are given $[Zn^{2+}]$, and so we solve the K_{sp} expression for the maximum $[S^{2-}]$ that can exist in the solution. Finally, we use the relationship for saturated H_2S solutions that contain a strong acid (Section 36-2) to calculate $[H^+]$.

$$[H^+]^2[S^{2-}] = 1.0 \times 10^{-27}$$

Solution

$$[Zn^{2+}][S^{2-}] = 1.1 \times 10^{-21}$$

$$[S^{2-}] = \frac{1.1 \times 10^{-21}}{[Zn^{2+}]} = \frac{1.1 \times 10^{-21}}{0.020} = 5.5 \times 10^{-20}\ M$$

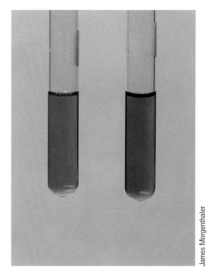

Copper(II) nitrate solutions are blue in excess HNO_3. Copper(II) chloride solutions are green in excess HCl.

Therefore, $[S^{2-}]$ cannot exceed $5.5 \times 10^{-20}\ M$ without exceeding K_{sp} for ZnS. In saturated H_2S solutions that contain a strong acid, the following relationship is valid. We can solve for $[H^+]$ because we know $[S^{2-}]$.

$$[H^+]^2[S^{2-}] = 1.0 \times 10^{-27}$$

$$[H^+]^2 = \frac{1.0 \times 10^{-27}}{[S^{2-}]} = \frac{1.0 \times 10^{-27}}{5.5 \times 10^{-20}} = 1.8 \times 10^{-8}$$

$$[H^+] = 1.3 \times 10^{-4}\ M$$

Thus, the concentration of H^+ must be at least $1.3 \times 10^{-4}\ M$ to prevent the precipitation of ZnS with the Groups I and II sulfides. The concentration of H^+ is adjusted to 0.3 M to provide a wide margin of safety.

Let us now demonstrate that cadmium ions are precipitated essentially completely from a solution under analytical Group I precipitation conditions.

EXAMPLE 31-2   *Equilibria in Sulfide Precipitations*

What is the maximum concentration of Cd^{2+} ions that can exist in a solution that is 0.30 M in H^+ and saturated with H_2S?

Plan

We know the relationship for saturated H_2S solutions that contain a strong acid.

$$[H^+]^2[S^{2-}] = 1.0 \times 10^{-27}$$

We solve for the maximum $[S^{2-}]$ possible in 0.30 M HCl solution. We know K_{sp} for CdS.

$$[Cd^{2+}][S^{2-}] = 3.6 \times 10^{-29}$$

Then we substitute the $[S^{2-}]$ into the K_{sp} expression for CdS to find the maximum $[Cd^{2+}]$ that can exist in this solution.

Solution

$$[H^+]^2[S^{2-}] = 1.0 \times 10^{-27}$$

$$[S^{2-}] = \frac{1.0 \times 10^{-27}}{[H]^{+2}} = \frac{1.0 \times 10^{-27}}{(0.30)^2} = 1.1 \times 10^{-26} \, M$$

The maximum concentration of sulfide ions is $1.1 \times 10^{-26} \, M$. Substitution of this value into the solubility product for cadmium sulfide gives

$$[Cd^{2+}][S^{2-}] = 3.4 \times 10^{-29}$$

$$[Cd^{2+}] = \frac{3.4 \times 10^{-29}}{[S^{2-}]} = \frac{3.4 \times 10^{-29}}{1.1 \times 10^{-26}} = \boxed{3.1 \times 10^{-3} \, M}$$

Thus, $3.1 \times 10^{-3} \, M$ is the maximum concentration of cadmium ions that can exist in a solution that is 0.30 M in H^+ and saturated with H_2S. These calculations show that most of the Cd^{2+} ions are precipitated under the Group I precipitation conditions.

31-5 MECHANISM OF SULFIDE PRECIPITATION

The concentration of sulfide ions in a solution saturated with H_2S and 0.3 M in H^+ ions is $1.1 \times 10^{-26} \, M$, or only about seven S^{2-} ions per 1000 L of solution (see Example 31-2). Yet the precipitation of many metal sulfides is essentially instantaneous in such solutions. This rapid precipitation of sulfides indicates that the reaction is *not* the combination of a simple metal ion and a simple sulfide ion. The concentration of hydrosulfide ions, HS^-, in saturated (0.10 M) H_2S solution that is also 0.3 M in HCl is much greater than the concentration of sulfide ions, as the following calculation shows.

Strictly speaking, HS^- is the hydrogen sulfide ion. However, it is commonly called *hydrosulfide ion*. Aqueous solutions of H_2S are usually called *hydrogen sulfide,* although they are properly named *hydrosulfuric acid.*

$$H_2S \rightleftharpoons H^+ + HS^- \qquad \frac{[H^+][HS^-]}{[H_2S]} = 1.0 \times 10^{-7}$$

$$[HS^-] = \frac{1.0 \times 10^{-7}[H_2S]}{[H^+]} = \frac{(1.0 \times 10^{-7})(0.10)}{0.30} = 3.3 \times 10^{-8} \, M$$

The concentration of the HS^- ions is about 10^{18} times the concentration of sulfide ions. We assume that a metal ion forms an unstable intermediate hydrosulfide salt that breaks down immediately to form the insoluble metal sulfide and H_2S. This is shown for the precipitation of CuS.

$$Cu^{2+}(aq) + 2HS^- \rightleftharpoons Cu(SH)_2(s) \rightleftharpoons CuS(s) + H_2S(aq)$$

The analogous formation of oxides, through precipitation of hydroxides followed by decomposition into oxides, is well known. For example, when hydroxide ions are added to solutions containing silver ions, an unstable intermediate, silver hydroxide, precipitates. Almost instantly the unstable silver hydroxide decomposes into silver oxide and water.

$$2Ag^+(aq) + 2OH^-(aq) \rightleftharpoons 2AgOH(s) \rightleftharpoons Ag_2O(s) + H_2O$$

Because equilibrium is concerned only with the final results of chemical reactions, and not with the mechanisms by which they occur, our calculations on the precipitation of sulfides are valid.

The equations for the precipitation of the Group I sulfides in a solution that is 0.30 M in HCl and saturated with H_2S may be written as

$$2Bi^{3+} + 3H_2S \longrightarrow Bi_2S_3(s) + 6H^+$$
dark brown

$$Cu^{2+} + H_2S \longrightarrow CuS(s) + 2H^+$$
black

$$Cd^{2+} + H_2S \longrightarrow CdS(s) + 2H^+$$
yellow

The sulfides of the Group I cations are all colored.

31-6 PRECIPITATION OF ANALYTICAL GROUP I CATIONS

Step A

Precipitation of the Group I sulfides. Use the known or unknown solution that contains some or all of Bi^{3+}, Cu^{2+}, and Cd^{2+}. To 10 drops of the known or unknown solution add 3 M aqueous ammonia until the solution is just basic to litmus. Now add 1 M HCl until the solution is just acidic to litmus. Add 2 drops of 6 M HCl and then dilute the solution to 1.5 mL. Mix well.

Add 12 drops of 5% thioacetamide solution. If directions have been followed carefully, the solution should now be approximately 0.3 M with respect to H^+. Heat the mixture in a hot water bath for 10 min to ensure hydrolysis of thioacetamide. Add 1 mL of water and continue to heat the mixture for 5 min.

The mixture should be stirred often during the heating process. Centrifuge and separate the precipitate from the solution. (For general unknowns *only*, save the centrifugate for the analysis of later groups. Add 2 drops of 12 M HCl to the centrifugate and place it in the hot water bath until all H_2S has been expelled. Wash the precipitate twice with 10 drops of 0.1 M HCl. The first 10 drops of wash solution should be added to the original centrifugate and the rest should be discarded.) The precipitate is treated according to Step B.

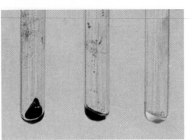

The sulfides of Group I. *Left to right:* Bi_2S_3 (*dark brown*), CuS (*black*), CdS (*yellow*).

31-7 DISSOLUTION OF ANALYTICAL GROUP I SULFIDES

Many metal sulfides become colloidal when they are washed with water. A solution of an electrolyte such as ammonium nitrate is used as the wash liquid to help prevent this.

The Group I sulfides are dissolved in hot dilute nitric acid, which oxidizes the sulfide ions to elemental sulfur. This decreases the concentration of sulfide ions so that the solubility products for the metal sulfides are no longer exceeded, $Q_{sp} < K_{sp}$, and so they dissolve. The equations for the dissolution of copper(II) sulfide are

$$CuS(s) \rightleftharpoons Cu^{2+} + S^{2-}$$
$$3S^{2-} + 8H^+ + 2NO_3^- \longrightarrow 3S(s) + 2NO(g) + 4H_2O$$

The equation for the net reaction is the sum of these two equations.

$$3CuS(s) + 8H^+ + 2NO_3^- \longrightarrow 3Cu^{2+} + 3S(s) + 2NO(g) + 4H_2O$$

The equations for the dissolution of the other Group I sulfides are similar. The reduction product from dilute nitric acid is nitric oxide, NO. Sulfur shows up in a variety of forms: sometimes as white flakes, sometimes in "powdery" forms, and sometimes as a "glob." In whatever form it appears, sulfur should be separated from the solution before beginning the next step.

Step B

Use the residue from Step A, containing CuS, CdS, and Bi_2S_3. Wash the solid sulfides with 16 drops of water to which 4 drops of 1 M NH_4NO_3 have been added, and discard the wash solution. Add 15 drops of 6 M HNO_3 to the residue, and heat the mixture in a water bath until the sulfides have dissolved. Stir while the mixture is being heated.

31-8 SEPARATION AND IDENTIFICATION OF BISMUTH(III) IONS

The hydroxides of bismuth(III), copper(II), and cadmium precipitate when NH_3 is added to the solution.

$$Bi^{3+} + 3NH_3 + 2H_2O \longrightarrow \underset{\text{white}}{Bi(OH)_3(s)} + 3NH_4^+$$

$$Cu^{2+} + 2NH_3 + 2H_2O \longrightarrow \underset{\text{light blue}}{Cu(OH)_2(s)} + 2NH_4^+$$

$$Cd^{2+} + 2NH_3 + 2H_2O \longrightarrow \underset{\text{white}}{Cd(OH)_2(s)} + 2NH_4^+$$

Bismuth(III) hydroxide is insoluble in (does not react with) excess aqueous NH_3. As excess aqueous ammonia is added, the hydroxides of copper(II) and cadmium dissolve by forming tetraammine complexes.

$$Cu(OH)_2(s) + 4NH_3 \rightleftharpoons \underset{\text{deep blue}}{[Cu(NH_3)_4]^{2+}} + 2OH^-$$

$$Cd(OH)_2(s) + 4NH_3 \rightleftharpoons \underset{\text{colorless}}{[Cd(NH_3)_4]^{2+}} + 2OH^-$$

Bismuth(III) hydroxide is separated from the soluble complexes of copper and cadmium. It is then treated with a powerful reducing agent, sodium stannite, $Na[Sn(OH)_3]$, in the presence of excess NaOH. Bismuth(III) ions are reduced to metallic bismuth, which is black, like most finely divided metals.

The systematic name for $[Sn(OH)_3]^-$ is trihydroxostannate(II) ion.

$$2Bi(OH)_3(s) + 3[Sn(OH)_3]^- + 3OH^- \longrightarrow 2Bi(s) + 3[Sn(OH)_6]^{2-}$$

Because sodium stannite is such a powerful reducing agent, it is not stable in the air. It is prepared just prior to use by treating a solution of tin(II) chloride dissolved in HCl with excess NaOH. The first few drops of NaOH produce a white precipitate, $Sn(OH)_2$, which dissolves in excess NaOH to produce the colorless stannite ion, $[Sn(OH)_3]^-$.

$$[SnCl_3]^- + 2OH^- \longrightarrow Sn(OH)_2(s) + 3Cl^-$$

$$Sn(OH)_2(s) + OH^- \longrightarrow [Sn(OH)_3]^-$$

The stannite ion disproportionates to give black or gray metallic tin on long exposure to air.

$$2[Sn(OH)_3]^- \longrightarrow Sn(s) + [Sn(OH)_6]^{2-}$$

Examine the solution of tin(II) chloride carefully before you prepare sodium stannite. The solution should be clear and there should be pieces of metallic tin in the bottom of the container to keep the tin reduced to the *+2 oxidation state.*

James Morgenthaler

The confirmatory test for bismuth before the mixture is stirred. Some white Bi(OH)$_3$ can be seen in the bottom of the test tube.

Step C

Use the solution from Step B, containing Bi^{3+}, Cu^{2+}, and Cd^{2+}. Add 15 *M* aqueous NH$_3$ dropwise until the solution is basic to litmus, and then add 5 more drops. Stir thoroughly. The appearance of a deep blue color shows the presence of copper as tetraamminecopper(II) ions.* The formation of a white precipitate indicates the presence of Bi^{3+} ions. Observe carefully because Bi(OH)$_3$ is a gelatinous precipitate that is difficult to see in the deep-blue-colored solution. Separate the mixture and save the solution for Step D. (Prepare sodium stannite by the dropwise addition of 4 *M* NaOH to 2 drops of a 1 *M* solution of tin(II) chloride until a white precipitate forms and then dissolves. Approximately 4–6 drops of NaOH will be required.) Add the freshly prepared sodium stannite solution to the white precipitate of Bi(OH)$_3$. Stir thoroughly. The *immediate* formation of a black solid, finely divided metallic bismuth, confirms the presence of bismuth(III) ions.

If the acidity is too low in unknowns that also contain Group III cations, the sulfides of zinc, nickel(II), and cobalt(II) will also precipitate. Nickel(II) ions follow copper(II) ions and give an ammine complex, $[Ni(NH_3)_6]^{2+}$, at this point. Its color is violet, quite different from that of $[Cu(NH_3)_4]^{2+}$.

31-9 IDENTIFICATION OF COPPER(II) AND CADMIUM IONS

The presence of copper(II) ions in the solution is confirmed by the deep blue color of the complex ion, $[Cu(NH_3)_4]^{2+}$. However, if only traces of copper are present, the blue color may be too faint to be seen. A portion of the solution is acidified with acetic acid, and some potassium hexacyanoferrate(II), also called potassium ferrocyanide, $K_4[Fe(CN)_6]$, is added. The formation of a deep-rose-colored precipitate, $Cu_2[Fe(CN)_6]$, is a very sensitive test for copper(II) ions. If copper is absent and cadmium is present, a white precipitate, $Cd_2[Fe(CN)_6]$, should be observed.

$$[Cu(NH_3)_4]^{2+} + 2OH^- + 6CH_3COOH \longrightarrow Cu^{2+} + 4NH_4^+ + 6CH_3COO^- + 2H_2O$$

$$[Cd(NH_3)_4]^{2+} + 2OH^- + 6CH_3COOH \longrightarrow Cd^{2+} + 4NH_4^+ + 6CH_3COO^- + 2H_2O$$

$$2Cu^{2+} + [Fe(CN)_6]^{4-} \longrightarrow Cu_2[Fe(CN)_6](s) \quad \text{deep rose}$$

$$2Cd^{2+} + Fe(CN)_6]^{4-} \longrightarrow Cd_2[Fe(CN)_6](s) \quad \text{white}$$

Nearly everything is contaminated with traces of iron. Iron(III) ions react with $[Fe(CN)_6]^{4-}$ ions to give a deep blue compound, $Fe_4[Fe(CN)_6]_3$. The "white" precipitate, $Cd_2[Fe(CN)_6]$, is usually very pale blue owing to traces of iron(III) ions. When copper is absent, this white precipitate serves as a confirmatory test for cadmium ions in Group I unknowns. However, when copper is present, a white precipitate cannot be observed in the presence of a deep-rose-colored precipitate. So an additional test for the presence of cadmium ions is necessary. (In general unknowns, students sometimes precipitate nickel sulfide in Group I. Nickel(II) ions follow copper(II) ions and give a pale blue precipitate, $Ni_2[Fe(CN)_6]$.)

When copper(II) ions are present, the presence of cadmium ions is confirmed by forming yellow CdS. However, before yellow CdS can be observed, copper(II) ions must be removed from the solution because they form a black sulfide, CuS. Two methods are used to detect cadmium ions in the presence of copper(II) ions. Each has some advantages and some disadvantages. Your instructor will indicate the preferred method.

In the first method, a portion of the basic solution that contains $[Cu(NH_3)_4]^{2+}$ and $[Cd(NH_3)_4]^{2+}$ ions is treated with $Na_2S_2O_4$, sodium dithionite, which reduces copper(II) to metallic copper.

$$[Cu(NH_3)_4]^{2+} + S_2O_4^{2-} + 2H_2O \longrightarrow Cu(s) + 2SO_3^{2-} + 4NH_4^+$$

However, it does not react with $[Cd(NH_3)_4]^{2+}$ ions. The addition of thioacetamide to the basic solution then results in the precipitation of CdS, a yellow compound, which confirms the presence of cadmium ions.

$$[Cd(NH_3)_4]^{2+} + S^{2-} \longrightarrow CdS(s) + 4NH_3$$

The advantage of the dithionite procedure is that traces of Bi(III) ions, which form a dark sulfide, are reduced to the free metal so that they do not interfere with the test for cadmium ions. The disadvantage of this procedure is that samples of sodium dithionite vary in composition. Some samples contain enough sodium sulfide, as an impurity, to form a (sulfide) precipitate as soon as the $Na_2S_2O_4$ is added. You should be alert to this possibility.

In the second method, also in ammoniacal solution, cyanide ions reduce copper(II) ions to copper(I) ions, which form very stable complex ions, $[Cu(CN)_2]^-$, that do not react with sulfide ions. Thus, the addition of excess potassium cyanide to the solution containing the tetraammine complexes of copper(II) and cadmium effectively removes the Cu^{2+} ions so that cadmium sulfide can be precipitated *and* observed.

$$2[Cu(NH_3)_4]^{2+} + 6CN^- \longrightarrow 2[Cu(CN)_2]^- + (CN)_2 + 8NH_3$$
$$\qquad\qquad\qquad\qquad\qquad\text{colorless}\qquad\text{cyanogen}$$

$$[Cu(CN)_2]^- + S^{2-} \longrightarrow \text{no reaction}$$

Cadmium ions also form complex ions in the presence of excess cyanide ions, but tetracyanocadmate ions are much less stable than dicyanocuprate(I) ions. They react with sulfide ions to form yellow cadmium sulfide.

$$[Cd(NH_3)_4]^{2+} + 4CN^- \longrightarrow [Cd(CN)_4]^{2-} + 4NH_3$$
$$\quad\text{colorless}\qquad\qquad\qquad\text{colorless}$$

$$[Cd(CN)_4]^{2-} + S^{2-} \longrightarrow \quad CdS(s) \quad + 4CN^-$$
$$\qquad\qquad\qquad\qquad\qquad\text{yellow}$$

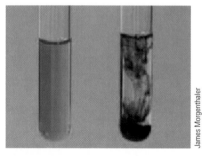

Confirmatory tests for copper. The blue solution contains $[Cu(NH_3)_4]^{2+}$ ions, and the deep-rose precipitate is $Cu_2[Fe(CN)_6]$.

James Morgenthaler

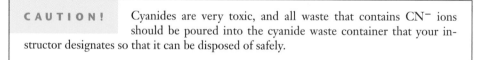

CAUTION!   Cyanides are very toxic, and all waste that contains CN^- ions should be poured into the cyanide waste container that your instructor designates so that it can be disposed of safely.

The advantage of the cyanide procedure is that it fairly effectively removes Ni^{2+} ions (a problem that occurs only in general unknowns when NiS has been precipitated in Group I rather than in Group III). The disadvantages are the toxicity of cyanides and the fact that any (stray) Bi^{3+} ions interfere with the Cd^{2+} test by forming a dark sulfide.

Step D

Use the solution from Step C, containing $[Cu(NH_3)_4]^{2+}$ and $[Cd(NH_3)_4]^{2+}$ ions. A trace of copper can be present even if the solution appears colorless. Place 10 drops of the solution that contains the tetraammine complexes of copper(II) and cadmium in a test tube, add 6 M CH_3COOH until the solution is acidic to litmus, and then add 3 drops of 0.2 M $K_4Fe(CN)_6$. A deep-rose-colored precipitate, $Cu_2[Fe(CN)_6]$, demonstrates the presence of copper. (This was already obvious from the fact that the ammoniacal solution was deep blue.) A pink precipitate, $Cu_2[Fe(CN)_6]$ possibly mixed with $Cd_2[Fe(CN)_6]$, indicates the presence of a very low concentration of Cu^{2+} ions. A white precipitate, $Cd_2[Fe(CN)_6]$, shows the presence of cadmium and the absence of copper. (If NiS has been precipitated in Group I, a pale blue-green precipitate, $Ni_2[Fe(CN)_6]$, may be observed here when copper(II) ions are absent.)

Use either of the following methods, (1) or (2), as directed by your instructor, to detect cadmium ions in the presence of copper(II) ions.

(1) To the remainder of the solution that contains $[Cu(NH_3)_4]^{2+}$ and $[Cd(NH_3)_4]^{2+}$, add an amount of $Na_2S_2O_4$ about the size of four grains of rice. Stir well and heat the mixture for 3 min. Centrifuge, and separate the dark precipitate of metallic copper. To the clear solution, add 3 drops of thioacetamide, stir well, and heat the mixture for 3 min. A yellow precipitate, CdS, confirms the presence of cadmium ions.

(2) Use the remainder of the solution containing the tetraammine complexes for the following test. In a hood, add 4 drops of 4 M KCN, mix thoroughly, and heat the solution for 1 min. Then add 2 drops of 5% thioacetamide solution. Heat the mixture in a hot water bath for 3 min. The formation of a yellow precipitate,* CdS, confirms the presence of cadmium. When Cu^{2+} ions are absent, the test for Cd^{2+} may be performed as described earlier by omitting the KCN treatment.

If your precipitate is so dark that you can't tell whether it contains CdS, the interfering dark sulfide Bi_2S_3 must be eliminated. Separate the mixture, discard the liquid into the cyanide waste container, and wash the precipitate twice with 10 drops of water to remove any adhering cyanide ions. Discard the wash liquid into the cyanide waste container. Add 6 drops of water *followed by* 3 drops of 6 M H_2SO_4 to the precipitate, and heat the mixture in the water bath for 2 min with constant stirring. Separate and discard any solid residue. Add 2 drops of 6 M aqueous ammonia to the clear, colorless liquid; mix well (the solution should be basic); and then add 2 drops of thioacetamide. Stir and heat for 3 min. The formation of a yellow precipitate confirms the presence of cadmium ions.

Traces of bismuth(III) ions interfere with the cadmium test by forming a dark sulfide. If traces of Bi^{3+} ions are present, the "yellow" sulfide precipitate may be olive to black depending on the amount of Bi^{3+} ions present. Because CdS is soluble in 2 M H_2SO_4 (Bi_2S_3 is not), it can easily be separated from the dark-colored sulfide.

Some alloys contain small percentages of copper.

Charles D. Winters

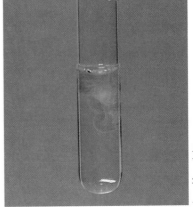

James Morgenthaler

The confirmatory tests for cadmium.

Exercises

General Questions on Analytical Group I

1. List the common oxidation states exhibited by the metals whose cations occur in Group I. Indicate those that are usually reducing oxidation states by (R), those that are considered "stable" oxidation states by (S), and those that are oxidizing by (O).

2. The known and unknown solutions contain 5.0 mg of *cation* per milliliter of solution. Calculate the molarity of each *in terms of the Group I metal*. For example, the $Cu(NO_3)_2$ solution contains 5.0 mg Cu/mL, and the molarity should be calculated in terms of mol Cu/L or mmol Cu/mL.

3. (a) What is thioacetamide? (b) For what is it used? Why? (c) Write the equation for the hydrolysis of thioacetamide in hot acidic solution.

4. What is the basis for the separation of the Group I cations from the later groups of cations?
5. Calculate the concentrations of HS^- and S^{2-} ions in saturated H_2S solution that is also 0.30 M in HCl (Section 36-2).
6. Calculate the concentration of S^{2-} in saturated H_2S solutions of (a) pH = 1.00, (b) pH = 2.00, and (c) pH = 3.00.
7. Copper(II) sulfide, $K_{sp} = 8.7 \times 10^{-36}$, is the least soluble of the Group I sulfides; cadmium sulfide, $K_{sp} = 3.4 \times 10^{-29}$, is much more soluble. (a) Calculate the concentrations of Cu^{2+} and Cd^{2+} ions that remain in solution, that is, unprecipitated, in a solution that is 0.30 M in HCl and saturated with H_2S. (b) What can you conclude about completeness of precipitation of these sulfides in this solution?
8. If the solution described in Exercise 7 were 0.15 M in HCl, what concentrations of Cu^{2+} and Cd^{2+} would remain unprecipitated?
9. (a) Based on your answers to Exercises 7 and 8, is there an obvious relationship between $[H^+]$ and $[Cu^{2+}]$ remaining in solution? What is it? (b) Is there a relationship between $[H^+]$ and $[Cd^{2+}]$ remaining in solution? What is it? (c) Is the relationship between $[H^+]$ and $[Cu^{2+}]$ similar to the relationship between $[H^+]$ and $[Cd^{2+}]$? Why? (d) Would the relationship between $[H^+]$ and $[Bi^{3+}]$ be of the same form? Why?
10. What color is each of the following? Aqueous solutions are indicated by (aq). $Cu(NO_3)_2$(aq); $Bi(NO_3)_3$(aq); $Cd(NO_3)_2$(aq); CuS; Bi_2S_3; CdS.
11. What color is each of the following? $Bi(OH)_3$; Bi; $Cu(OH)_2$; $Cu(NH_3)_4(OH)_2$(aq); $Cu_2[Fe(CN)_6]$; $Cd(OH)_2$; $Cd(NH_3)_4(OH)_2$(aq); $Cd_2[Fe(CN)_6]$.

Analytical Group I Reactions

Write balanced net ionic equations for the reactions that occur when the following substances are mixed in aqueous solution. Indicate the colors of all precipitates and complex ions.

Step A

12. Bismuth(III) chloride and hydrogen sulfide.
13. Copper(II) chloride and hydrogen sulfide.
14. Cadmium chloride and hydrogen sulfide.

Step B

15. Solid bismuth(III) sulfide and hot 6 M nitric acid.
16. Solid copper(II) sulfide and hot 6 M nitric acid.
17. Solid cadmium sulfide and hot 6 M nitric acid.

Step C

18. Bismuth(III) sulfate and concentrated aqueous ammonia.
19. Copper(II) sulfate and a limited amount of concentrated aqueous ammonia.
20. Copper(II) hydroxide and excess concentrated aqueous ammonia.

21. Cadmium sulfate and a limited amount of concentrated aqueous ammonia.
22. Cadmium hydroxide and excess concentrated aqueous ammonia.
23. Tin(II) chloride in HCl and a limited amount of NaOH.
24. Solid tin(II) hydroxide and excess NaOH.
25. Solid bismuth(III) hydroxide and sodium stannite (in excess NaOH).

Step D

26. Tetraamminecopper(II) hydroxide and acetic acid.
27. Tetraamminecadmium hydroxide and acetic acid.
28. Copper(II) acetate and potassium hexacyanoferrate(II).
29. Cadmium acetate and potassium hexacyanoferrate(II).
30. Tetraamminecopper(II) hydroxide and excess KCN.
31. Tetraamminecadmium hydroxide and excess KCN.
32. Potassium tetracyanocadmate and $(NH_4)_2S$.
33. Tetraamminecopper(II) hydroxide and sodium dithionite.
34. Tetraamminecadmium hydroxide and $(NH_4)_2S$.

Other Questions and Problems

35. Explain the equilibria involved in the dissolution of the Group I sulfides in hot 6 M HNO_3.
36. (a) Which of the Group I cations form hydroxides that are soluble in excess aqueous ammonia? (b) What are the formulas for their ammine complexes?
37. Which of the Group I cations form hydroxides that are soluble in excess NaOH solution?
38. Why is sodium stannite prepared just before it is used in the confirmatory test for bismuth(III) ions?
39. In Step D the ammine complexes of Cu^{2+} and Cd^{2+} are destroyed by adding CH_3COOH before $[Fe(CN)_6]^{4-}$ ions are added. Why?
40. A solution is 0.095 M each in $Cu(NO_3)_2$, $Bi(NO_3)_3$, $Cd(NO_3)_2$, and HCl. If sufficient H_2S is added so that the solution is saturated when precipitation is as complete as possible, what concentrations of Cu^{2+}, Bi^{3+}, and Cd^{2+} remain in solution?
41. Is it possible to increase the acidity of a solution to the point that CdS will not precipitate if the solution is 0.050 M in $Cd(NO_3)_2$ and saturated with H_2S? Justify your answer by appropriate calculations.
42. Aqueous ammonia was added to a Group I unknown. A precipitate formed initially. All of it dissolved and the solution turned deep blue as more aqueous ammonia was added. Which cation(s) could be present?
43. No precipitate was observed in a Group I unknown that was also 6 M in aqueous ammonia. Which cations could be present?
44. A Group I unknown gave a yellow sulfide precipitate in the initial precipitation of the group. Which cations could be present?
45. A Group I unknown gave a black sulfide precipitate in the initial precipitation of the group. Which cations could be present?

Analysis of Cation Group II

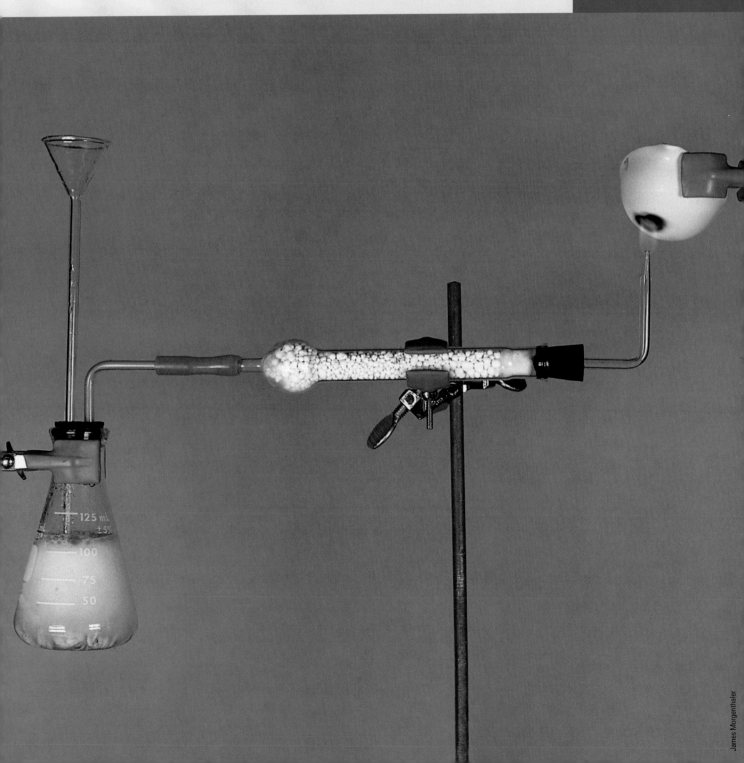

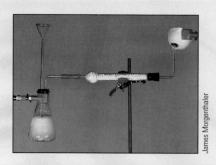

This photo shows a modification of the Marsh test for arsenic, a classic procedure in criminology labs. A solution containing arsenic was added to the flask, which contains Zn and dilute H_2SO_4. As gaseous arsine, AsH_3, is formed, it is carried in the stream of H_2 produced by the reaction of Zn and H_2SO_4. Anhydrous $CaCl_2$ absorbs H_2O vapor in the H_2/AsH_3 gas stream. Hydrogen is ignited as it escapes from the glass tube at the right. Arsine imparts a yellow-green color to the nearly colorless H_2/O_2 flame. The dark spot on the bottom of the evaporating dish is elemental arsenic.

James Morgenthaler

OUTLINE

32-1 COMMON OXIDATION STATES OF METALS IN CATION GROUP II

Arsenic is classified as a metalloid. As expected, it exhibits some properties of both metals and nonmetals. In aqueous solutions its important oxidation states are +3 and +5. It can be reduced to the −3 oxidation state by strong reducing agents. As the following half-reaction indicates, arsenic(V) is a mild oxidizing agent.

$$H_3AsO_4(aq) + 2H^+(aq) + 2e^- \longrightarrow H_3AsO_3(aq) + H_2O \qquad E^0 = +0.58 \text{ V}$$

Antimony is decidedly more metallic than arsenic. Compounds containing antimony in the +5 oxidation state are stronger oxidizing agents than are similar compounds of arsenic.

$$[SbCl_6]^-(aq) + 2e^- \longrightarrow [SbCl_4]^-(aq) + 2Cl^-(aq) \qquad E^0 = +0.75 \text{ V}$$

In aqueous solutions the most important oxidation state of Sb is +3.

Figure 32-1 Analytical Group II flow chart. Circled letters refer to analytical steps. Species formed in confirmatory tests are shown with blue backgrounds. Precipitated species are underlined.

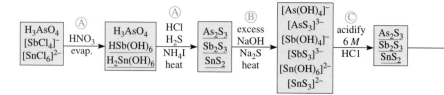

Antimony and arsenic, two metalloids, are both semiconducting elements. A bar of antimony (*left*) with a sample of one of its ores, stibnite; arsenic (*right*).

Tin exhibits two well-defined oxidation states, Sn(II) and Sn(IV). In the +2 oxidation state, tin is a mild reducing agent in acidic solution.

$$Sn^{4+}(aq) + 2e^- \longrightarrow Sn^{2+}(aq) \qquad E^0 = +0.15 \text{ V}$$

Metallic tin is added to aqueous solutions of Sn(II) salts to prevent atmospheric oxidation. In basic solutions Sn(II) is such a strong reducing agent that it is unstable.

32-2 INTRODUCTION TO THE ANALYTICAL PROCEDURES

Like the cations of analytical Group I, the Group II cations form sulfides that are insoluble in dilute acidic solutions. However, sulfides of the Group II cations are soluble in solutions of strong soluble bases, and so the cations of analytical Group II are referred to as the **acid-insoluble, base-soluble sulfide group.** Group II sulfides are quite insoluble in water (Table 32–1).

Exact values of the solubility products for arsenic(III) sulfide, As_2S_3; antimony(III) sulfide, Sb_2S_3; and tin(IV) sulfide, SnS_2, are difficult to determine experimentally. This is because both the cations and the anions of these sulfides hydrolyze extensively in complex reactions. However, many experiments have demonstrated conclusively that these sulfides are quite insoluble.

The flow chart in Figure 32–1 outlines schematically the precipitation of the Group II sulfides and the analysis of this group of cations.

TABLE 32-1	*Solubility Products for Some Group II Sulfides*
Sulfide	K_{sp}
Sb_2S_3	1.6×10^{-93}
SnS_2	1×10^{-70}

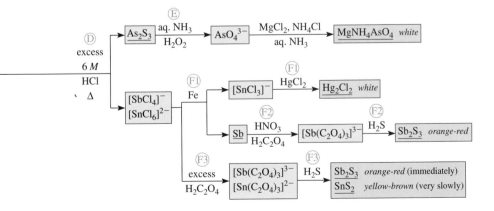

32-3 PREPARATION OF SOLUTIONS OF ANALYTICAL GROUP II CATIONS

Solutions of antimony and tin are usually prepared by dissolving their chlorides in dilute hydrochloric acid to prevent hydrolysis of the metal ions.

$$SbCl_3(s) + [H^+ + Cl^-] \longrightarrow [H^+ + SbCl_4^-] \qquad \text{tetrachloroantimonic(III) acid}$$

$$SnCl_4(\ell) + 2[H^+ + Cl^-] \longrightarrow [2H^+ + SnCl_6^{2-}] \qquad \text{hexachlorostannic(IV) acid}$$

covalent
compounds   chloro acids
                                            (strong acids)

Solutions containing arsenic are conveniently prepared by dissolving diarsenic pentoxide in hot water.

$$As_2O_5(s) + 3H_2O \xrightarrow{\text{heat}} 2H_3AsO_4(aq) \qquad \text{arsenic acid (a weak acid)}$$

32-4 EVAPORATION WITH NITRIC ACID

Before the analytical Group II precipitation takes place, the solution is "cleaned up" by evaporation with nitric acid. This converts antimony and tin to insoluble acids in their highest oxidation states, Sb(V) and Sn(IV). Arsenic is already in its highest oxidation state, As(V), in H_3AsO_4.

The formula for $HSb(OH)_6$ has been established. The formula for $H_2Sn(OH)_6$ has not. It is a hydrated oxide, $SnO_2 \cdot xH_2O$, which is written as either H_2SnO_3 or $H_2Sn(OH)_6$. Salts derived from this compound are of the type $Na_2Sn(OH)_6$, so we will represent the acid, or hydrated oxide, as $H_2Sn(OH)_6$.

$$3[SbCl_4]^- + 5H^+ + 2NO_3^- + 14H_2O \xrightarrow[\text{HNO}_3]{\text{evap.}} 3HSb(OH)_6(s) + 2NO + 12HCl(g)$$

antimonic acid
white

$$[SnCl_6]^{2-} + 2H^+ + 6H_2O \xrightarrow[\text{HNO}_3]{\text{evap.}} H_2Sn(OH)_6(s) + 6HCl(g)$$

stannic acid
white

The precipitation of arsenic(V) sulfide is very slow. Ammonium iodide, NH_4I, is added to reduce arsenic from the +5 to the +3 oxidation state so that it can be precipitated more rapidly as arsenic(III) sulfide.

$$H_3AsO_4(aq) + 2H^+(aq) + 2I^-(aq) \longrightarrow H_3AsO_3(s) + I_2(s) + H_2O$$

arsenic acid   arsenous acid

Antimony is also reduced to the +3 oxidation state by iodide ions.

$$HSb(OH)_6(s) + 2H^+(aq) + 2I^-(aq) \longrightarrow H_3SbO_3(s) + I_2(s) + 3H_2O$$

antimonic acid   antimonous acid

Iodine oxidizes H_2S to free sulfur and reappears as iodide ions in the solution.

$$I_2(s) + H_2S(aq) \longrightarrow 2H^+(aq) + 2I^-(aq) + S(s)$$

32-5 PRECIPITATION OF ANALYTICAL GROUP II CATIONS

The solution containing analytical Group II cations is subjected to a preliminary treatment, that is, evaporation with HNO_3 and then reaction with NH_4I. However, the precipitation conditions are the same as for analytical Group I cations, that is, the solution of 0.3 M in H^+ and saturated with H_2S. The equations for the precipitation of the Group II sulfides in

this solution may be written as

$$2H_3AsO_3(s) + 3H_2S \longrightarrow As_2S_3(s) + 6H_2O$$
yellow

$$2H_3SbO_3(s) + 3H_2S \longrightarrow Sb_2S_3(s) + 6H_2O$$
orange-red

$$H_2Sn(OH)_6(s) + 2H_2S \longrightarrow SnS_2(s) + 6H_2O$$
light tan

The sulfides of the Group II cations are colored.

H_3AsO_3 and H_3SbO_3 may also be written as $As(OH)_3$ and $Sb(OH)_3$ because both are amphoteric. They are also written as $HAsO_2$ and $HSbO_2$, which contain one less H_2O per formula unit.

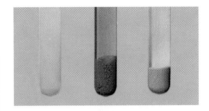

The sulfides of Group II. Left to right: As_2S_3 (*pale yellow*), Sb_2S_3 (*orange-red*), SnS_2 (*light tan*).

Step A

Precipitation of the analytical Group II sulfides. Use the known or unknown solution that contains some or all of H_3AsO_4, $SbCl_4^-$, and $SnCl_6^{2-}$. Add 4 drops of 6 M HNO_3 to 10 drops of the known or unknown solution and evaporate the solution to a moist residue (Section 30-16). Care should be taken to avoid evaporating the precipitate to dryness. Cool the moist residue and add 10 drops of water, mix well, and add 1 M aqueous ammonia until the solution is just basic to litmus. Now add 1 M HCl until the solution is just acidic to litmus. Add 2 drops of 6 M HCl and 1 drop of 1 M NH_4I, and then dilute the solution to 1.5 mL. Mix well.

Heat the solution in the water bath for 3 min. Add 12 drops of 5% thioacetamide solution. If directions have been followed carefully, the solution should now be approximately 0.3 M with respect to H^+. Heat the mixture in a hot water bath for 10 min to ensure hydrolysis of thioacetamide. Add 1 mL of water and continue to heat the mixture for 5 min.

The mixture should be stirred often during the heating process. Centrifuge and separate the precipitate from the solution. (For general unknowns *only*, save the centrifugate for the analysis of Groups III to V. Add 2 drops of 12 M HCl to the centrifugate and place it in the hot water bath until all H_2S has been expelled. Wash the precipitate twice with 10 drops of 0.1 M HCl. The first 10 drops of wash solution should be added to the original centrifugate, and the rest should be discarded.) The precipitate is treated according to Step B.

32-6 DISSOLUTION OF ANALYTICAL GROUP II SULFIDES IN SODIUM HYDROXIDE SOLUTION

The hydroxides of arsenic(III), antimony(III), and tin(IV) are amphoteric (Table 29-3). Their sulfides exhibit **thioamphoterism;** that is, the sulfides of these three elements form soluble complex compounds in strongly basic solutions. Sodium hydroxide (NaOH) is the reagent used to dissolve these sulfides. A little thioacetamide is also added so there will be plenty of sulfide ions when we are ready to reprecipitate these sulfides.

$$As_2S_3(s) + 4OH^- \longrightarrow [As(OH)_4]^- + [AsS_3]^{3-} \qquad \text{trithioarsenate(III) ion[1]}$$

---

[1] *You may wonder whether species such as $[As(OH)_2S]^-$ and $[As(OH)S_2]^{2-}$ are formed in this reaction. Yes, they are. Similar species are also formed in the dissolution reactions of Sb_2S_3 and SnS_2. We have represented the reactions in the simplest way that is consistent with accuracy.*

$$Sb_2S_3(s) + 4OH^- \longrightarrow [Sb(OH)_4]^- + [SbS_3]^{3-} \qquad \text{trithioantimonate(III) ion}$$

$$3SnS_2(s) + 6OH^- \longrightarrow [Sn(OH)_6]^{2-} + 2[SnS_3]^{2-} \qquad \text{trithiostannate(IV) ion}$$

The prefix *thio-* always refers to sulfur.

The sulfur-containing products of these reactions are called *thio* salts. (The Na^+ ion is the only cation present in significant concentrations in this strongly basic solution.)

C A U T I O N !

Failure to dissolve the Group II sulfides in the NaOH solution results in arsenic, antimony, and tin being missed.

As_2S_3, Sb_2S_3, and SnS_2 are soluble in 4 M NaOH.

Step B
Use the precipitate from Step A, containing As_2S_3, Sb_2S_3, and SnS_2. Add 12 drops of 4 M NaOH and 5 drops of thioacetamide solution to the Group II sulfide precipitate. Heat the mixture in a water bath for 5 min, stirring constantly. The precipitate will dissolve completely for unknowns that contain only Group II cations. A precipitate will remain in unknowns that contain *both* Groups I and II. Centrifuge the mixture and treat the clear solution according to Step C. Use the precipitate (Group I sulfides) to begin the analysis for Group I cations with Step B in Group I.

After we have completed the analysis of Group II cations, we shall analyze an unknown solution that contains ions from both Group I and Group II. We use this procedure (Step B), dissolution of Group II sulfides in sodium hydroxide solution, to separate Group I and Group II cations. The sulfides of Group I cations are insoluble in sodium hydroxide solution.

32-7 REPRECIPITATION OF ANALYTICAL GROUP II SULFIDES

Group II sulfides can be reprecipitated from the strongly basic solution of their hydroxo-complexes and thiosalts by removing the excess S^{2-} and OH^- ions. The addition of an acid decreases the basicity of the solution and destroys the hydroxocomplexes and thiosalts. Sulfide ions then combine with Group II cations to reprecipitate their insoluble sulfides. The equations for the equilibria for the reprecipitation of tin(IV) sulfide from its thiosalt are

$$[SnS_3]^{2-} \rightleftharpoons SnS_2(s) + S^{2-}$$
$$\underline{S^{2-} + 2H^+ \longrightarrow H_2S(g)}$$
net reaction $\quad [SnS_3]^{2-} + 2H^+ \longrightarrow SnS_2(s) + H_2S(g)$

The equilibria for the precipitation of SnS_2 from its hydroxocomplex are

$$[Sn(OH)_6]^{2-} + 6H^+ \longrightarrow Sn^{4+} + 6H_2O$$
$$\underline{Sn^{4+} + 2H_2S \longrightarrow SnS_2(s) + 4H^+}$$
net reaction $\quad [Sn(OH)_6]^{2-} + 2H^+ + 2H_2S \longrightarrow SnS_2(s) + 6H_2O$

These equations show that the addition of a strong acid displaces the equilibrium in the direction of insoluble tin(IV) sulfide, $[Sn^{4+}][S^{2-}]^2 > K_{sp}$, because of the formation of an insoluble compound, SnS_2, and a weak electrolyte, H_2O. Multiplying the first equation by two and combining with the second equation gives

$$2[SnS_3]^{2-} + [Sn(OH)_6]^{2-} + 6H^+ \longrightarrow 3SnS_2(s) + 6H_2O$$

The net ionic equations for the reprecipitation of the Group II sulfides are

$$[AsS_3]^{3-} + [As(OH)_4]^- + 4H^+ \longrightarrow As_2S_3(s) + 4H_2O$$

$$[SbS_3]^{3-} + [Sb(OH)_4]^- + 4H^+ \longrightarrow Sb_2S_3(s) + 4H_2O$$

$$2[SnS_3]^{2-} + [Sn(OH)_6]^{2-} + 6H^+ \longrightarrow 3SnS_2(s) + 6H_2O$$

Failure to acidify the solution that contains the hydroxocomplexes and the thiosalts of the Group II cations is a common mistake. If the solution is not acidified properly, the Group II sulfides fail to reprecipitate, and the ions are discarded. Ions poured into the waste container will be missed!

Recall from Step B that $[SnS_3]^{2-}$ and $[Sn(OH)_6]^{2-}$ are formed in a 2:1 mole ratio.

Step C

Use the solution from Step B, containing hydroxocomplexes and thiosalts of the Group II ions: $[AsS_3]^{3-}$, $[As(OH)_4]^-$, $[SbS_3]^{3-}$, $[Sb(OH)_4]^-$, $[SnS_3]^{2-}$, and $[Sn(OH)_6]^{2-}$ ions. Add 6 M HCl, with vigorous stirring, until the solution is just acidic to litmus. Heat the mixture in a water bath for 5 min. (Because large amounts of H_2S are evolved as the solid sulfides form in the solution, caution must be exercised to determine that the solution is acidic throughout. When you are convinced that the solution is acidic, stir it thoroughly. Then test for acidity again.) Separate the mixture, discard the solution, and save the precipitate of As_2S_3, Sb_2S_3, and SnS_2 for Step D.

32-8 SEPARATION OF ARSENIC FROM ANTIMONY AND TIN

The K_{sp}'s for the Group II sulfides are quite small. To dissolve Sb_2S_3 and SnS_2, while leaving solid As_2S_3 behind, the concentrations of H^+ and S^{2-} ions must be adjusted so that $[Sb^{3+}]^2[S^{2-}]^3 < K_{sp}$ and $[Sn^{4+}][S^{2-}]^2 < K_{sp}$, while $[As^{3+}]^2[S^{2-}]^3 > K_{sp}$. Hot 6 M HCl reduces the concentrations of the ions of antimony(III) and tin(IV) sulfides to values such that these sulfides dissolve.

$$SnS_2(s) + 4H^+ + 6Cl^- \rightleftharpoons [SnCl_6]^{2-} + 2H_2S(g)$$

$$Sb_2S_3(s) + 6H^+ + 8Cl^- \rightleftharpoons 2[SbCl_4]^- + 3H_2S(g)$$

The sulfide of arsenic is not dissolved by 6 M HCl because its solubility product constant is so small. The reactions in which Sb_2S_3 and SnS_2 dissolve in hot 6 M HCl are *reversible*. Reprecipitation of Sb_2S_3 and SnS_2 may occur when the solution cools. If the solution turns orange or yellow *after* it has been drawn into the capillary pipet, the separation has been accomplished. If it turns orange or yellow *before* it is separated from As_2S_3, the mixture must be heated again before separation is possible.

Step D

Use the precipitate from Step C, containing As_2S_3, Sb_2S_3, and SnS_2. Add 1 mL of 6 M HCl to the precipitate, stir the mixture thoroughly, and heat it in a hot water bath for 3 min with regular stirring. Separate the residue *as quickly as possible*, and save the centrifugate. Add 1 mL of 6 M HCl to the residue and heat the mixture for 3 min, stirring regularly. Separate any residue, and save the combined centrifugates for Step F. (If the HCl solution becomes cool, Sb_2S_3 and SnS_2 may reprecipitate. If the solution turns orange or yellow *inside* the capillary pipet, the separation has been accomplished and there is no need to worry.) The residue is analyzed by Step E. Place the solution containing $[SbCl_4]^-$ and $[SnCl_6]^{2-}$ in the hot water bath so that H_2S will be expelled from it while you are identifying arsenic.

32-9 IDENTIFICATION OF ARSENIC

A saturated solution of arsenic(III) sulfide contains such a low concentration of S^{2-} ions that this sulfide cannot be dissolved completely by the addition of H^+ (from strong acids). The product of $[H^+]^2$ and $[S^{2-}]$ does not exceed the ion product constant for H_2S, even in solutions containing the maximum $[H^+]$ attainable, because $[S^{2-}]$ is so small. Therefore, equilibrium is established before arsenic(III) sulfide dissolves completely.

However, the concentration of sulfide ions can be reduced by *oxidation* so that As_2S_3 dissolves. A mixture of aqueous ammonia and hydrogen peroxide dissolves As_2S_3.

$$As_2S_3(s) + 14H_2O_2 + 12OH^- \longrightarrow 2AsO_4^{3-} + 3SO_4^{2-} + 20H_2O$$

Arsenic(III) ions are oxidized to arsenate ions, AsO_4^{3-}, and sulfide ions are oxidized to sulfate ions. The concentrations of both As^{3+} and S^{2-} are lowered to the point that $[As^{3+}]^2[S^{2-}]^3 < K_{sp}$, so As_2S_3 dissolves.

When magnesia mixture is added to a solution containing arsenate ions, a white crystalline compound, magnesium ammonium arsenate, $MgNH_4AsO_4$, precipitates. Magnesia mixture is a buffered aqueous ammonia–ammonium chloride solution that contains magnesium chloride.

$$Mg^{2+} + NH_4^+ + AsO_4^{3-} \longrightarrow MgNH_4AsO_4(s) \qquad \text{white}$$

The confirmatory test for arsenic, $MgNH_4AsO_4$.

Step E

Use the residue from Step D, As_2S_3. Add 8 drops of 6 M aqueous ammonia, 4 drops of water, and 6 drops of 6% hydrogen peroxide to the As_2S_3. Stir thoroughly and heat the mixture in a hot water bath with constant stirring for 5 min. Separate the mixture. If the solution containing AsO_4^{3-} ions also contains colloidal sulfur, wrap a small piece of cotton around the tip of a capillary pipet and then draw the liquid into the pipet. Most of the colloidal sulfur should be trapped by the cotton. Transfer the liquid to a clean test tube and add 2 drops of 15 M aqueous ammonia and 5 drops of magnesia mixture to the solution. The formation of a white precipitate, $MgNH_4AsO_4$, confirms the presence of arsenate ions. The precipitate may form slowly. Scratching the test tube walls with a stirring rod usually hastens crystallization.

32-10 IDENTIFICATION OF ANTIMONY AND TIN

Recall that antimony(III) and tin(IV) sulfides were dissolved in hot $6\,M$ HCl and are in the form of complex ions, $[SbCl_4]^-$ and $[SnCl_6]^{2-}$. Active metals reduce antimony to the metallic state and tin to the $+2$ oxidation state in hydrochloric acid solution. This reaction provides a convenient method for separating antimony and tin. We use a small iron nail as a source of an active metal.

$$2[SbCl_4]^- + 3Fe(s) \longrightarrow 2Sb(s) + 3Fe^{2+} + 8Cl^-$$
$$[SnCl_6]^{2-} + Fe(s) \longrightarrow [SnCl_3]^- + Fe^{2+} + 3Cl^-$$

Tin is a reducing agent in the $+2$ oxidation state. The reaction of tin(II) chloride with mercury(II) chloride is used as the confirmatory test for tin. The reactions occur in a solution that contains an excess of hydrochloric acid.

$$[SnCl_3]^- + 2[HgCl_4]^{2-} \longrightarrow Hg_2Cl_2(s) + [SnCl_6]^{2-} + 3Cl^-$$
$$Hg_2Cl_2(s) + [SnCl_3]^- + Cl^- \longrightarrow 2Hg(\ell) + [SnCl_6]^{2-}$$

Hg_2Cl_2 is a white, highly crystalline substance as it is formed in this reaction. A little of it is reduced to liquid mercury by $[SnCl_3]^-$ ions.

If black flakes of metallic antimony are formed when the nail is placed in the solution, this provides convincing evidence that antimony was present in the solution. However, we shall perform a confirmatory test. Antimony reacts with dilute nitric acid to form the insoluble oxide, Sb_4O_6. This oxide dissolves in oxalic acid, $H_2C_2O_4$, to form a soluble complex compound, $H_3[Sb(C_2O_4)_3]$.

$$4Sb(s) + 4H^+ + 4NO_3^- \longrightarrow Sb_4O_6(s) + 4NO(g) + 2H_2O$$
$$Sb_4O_6(s) + 12H_2C_2O_4 \longrightarrow 12H^+ + 4[Sb(C_2O_4)_3]^{3-} + 6H_2O$$

These two equations are frequently combined into a single equation

$$Sb(s) + NO_3^- + 3H_2C_2O_4 \longrightarrow [Sb(C_2O_4)_3]^{3-} + 2H^+ + NO(g) + 2H_2O$$

The precipitation of the orange-red sulfide, Sb_2S_3, is the confirmatory test.

$$2[Sb(C_2O_4)_3]^{3-} + 3H_2S + 6H^+ \longrightarrow Sb_2S_3(s) + 6H_2C_2O_4$$

Excess nitric acid remaining in the solution oxidizes hydrogen sulfide to free sulfur and interferes with the precipitation of Sb_2S_3. Therefore, it is important (a) to add no more nitric acid than necessary to dissolve antimony and (b) to add an excess of thioacetamide.

An alternative test for antimony(III) is available. Both Sb(III) and Sn(IV) ions react with oxalic acid to form soluble complex compounds.

$$[SbCl_4]^- + 3H_2C_2O_4 \longrightarrow [Sb(C_2O_4)_3]^{3-} + 6H^+ + 4Cl^-$$
$$[SnCl_6]^{2-} + 3H_2C_2O_4 \longrightarrow [Sn(C_2O_4)_3]^{2-} + 6H^+ + 6Cl^-$$

However, the trisoxalatostannate(IV) ion is much more stable than the trisoxalatoantimonate(III) ion. Its stability allows us to detect antimony(III) ions in the presence of Sn(IV) ions. The less stable $[Sb(C_2O_4)_3]^{3-}$ ion reacts with H_2S immediately to give an orange-red precipitate, Sb_2S_3. The $[Sn(C_2O_4)_3]^{2-}$ ion is so stable that it reacts with H_2S only *very slowly* to produce SnS_2, which is light tan.

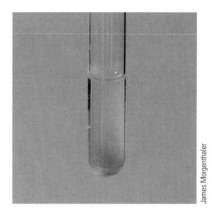

The confirmatory test for tin.

James Morgenthaler

Oxalic acid is sometimes written as $(COOH)_2$,

$[Sn(C_2O_4)_3]^{2-}$ is a very stable complex ion.

The confirmatory test for antimony, Sb_2S_3.

Step F

Use the solution from Step D, containing $[SbCl_4]^-$ and $[SnCl_6]^{2-}$ ions. All the H_2S should have been expelled by boiling this solution while the test for arsenic was performed.

Perform either (1) and (2) only or (1), (2), and (3) as directed by your instructor. If you perform only (1) and (2), use all of the solution that contains $[SbCl_4]^-$ and $[SnCl_6]^{2-}$ ions for (1). If you perform (1), (2), and (3), use one half of the solution for (1) and one half for (3).

(1) Add 2 drops of 6 M HCl to (*one half of*) the solution. Place a clean, small iron nail in the test tube and heat the mixture for 5 min. Black flakes of metallic antimony should appear. Withdraw the liquid and place it in a clean test tube. Add 2 drops of 6 M HCl to the solution and heat it in a water bath for 3 min. Add 5 drops of 0.2 M $HgCl_2$ to the warm solution. The formation of a white or gray precipitate confirms the presence of tin.

(2) Remove the iron nail, leaving the black flakes of metallic antimony in the test tube. (Small black pieces of carbon may remain if a significant amount of the nail dissolved. These are usually very small pieces that do not resemble flakes of antimony.) Place the nail in a solid disposal receptacle. Add 1 drop of 6 M HNO_3 and an amount of solid $H_2C_2O_4 \cdot 2H_2O$ about the size of a pea to the black flakes. Warm the mixture in a water bath until the black flakes dissolve. (If the black flakes fail to dissolve in 3 min, add 1 more drop of 6 M HNO_3 and warm the solution.) Add 10 drops of H_2O and 5 drops of thioacetamide to the solution. Stir thoroughly and then place the test tube in a hot water bath for 3 min. The formation of an orange-red precipitate, Sb_2S_3, confirms the presence of antimony. (Excess nitric acid oxidizes hydrogen sulfide to free sulfur. Do not confuse free sulfur with an orange-red precipitate of Sb_2S_3.)

(3) *Alternative test for antimony:* Add an amount of solid oxalic acid dihydrate, $H_2C_2O_4 \cdot 2H_2O$, about the size of a pea to *one half* of the solution that contains $[SbCl_4]^-$ and $[SnCl_6]^{2-}$ ions. Place the test tube in a hot water bath and stir it for 3 min. Some solid $H_2C_2O_4 \cdot 2H_2O$ should remain—if not, add more and heat for an additional 3 min. Separate the clear, colorless liquid and place it in a clean test tube. Add 3 drops of thioacetamide, mix well, and place the test tube in the hot water bath for 3 min. The formation of an orange-red precipitate, Sb_2S_3, confirms the presence of antimony. A light tan precipitate that forms very slowly (after a few minutes) is probably tin(IV) sulfide.

Exercises

General Questions on Analytical Group II

1. List the common oxidation states exhibited by the metals whose cations occur in Group II. Indicate those that are usually reducing oxidation states by (R), those that are considered "stable" oxidation states by (S), and those that are oxidizing by (O).

2. The known and unknown solutions contain 5.0 mg of *cation* per milliliter of solution. Calculate the molarity of each *in terms of the Group II metal.* For example, the $HSbCl_4$ solution contains 5.0 mg Sb/mL, and the molarity should be calculated in terms of mol Sb/L or mmol Sb/mL.

3. What color is each of the following? H_3AsO_4(aq); $H[SbCl_4]$(aq); $H_2[SnCl_6]$(aq); As_2S_3; Sb_2S_3; $MgNH_4AsO_4$.

Analytical Group II Reactions

Write balanced net ionic equations for the reactions that occur when the following substances are mixed in aqueous solution. Indicate the colors of all precipitates and complex ions.

Solution Preparation

4. Antimony(III) chloride dissolves in dilute HCl.
5. Tin(IV) chloride dissolves in dilute HCl.
6. Arsenic(V) oxide dissolves in hot water.

Step A

7. Antimony(III) chloride in dilute HCl and evaporation with HNO_3.
8. Tin(IV) chloride in dilute HCl and evaporation with HNO_3.
9. Arsenic acid and ammonium iodide in excess HCl.
10. Antimonic acid and ammonium iodide in excess HCl.
11. Arsenous acid and hydrogen sulfide.
12. Antimonous acid and hydrogen sulfide.
13. Stannic acid and hydrogen sulfide.

Step B

14. Arsenic(III) sulfide in excess NaOH.
15. Antimony(III) sulfide in excess NaOH.
16. Tin(IV) sulfide in excess NaOH.

Step C

17. Sodium trithioarsenate(III) and sodium tetrahydroxoarsenate(II) plus a limited amount of dilute HCl.
18. Sodium trithioantimonate(III) and sodium tetrahydroxoantimonate(III) plus a limited amount of dilute HCl.
19. Sodium trithiostannate(IV) and sodium hexahydroxostannate(IV) plus a limited amount of dilute HCl.

Step D

20. Tin(IV) sulfide and excess hot $6\,M$ HCl.
21. Antimony(III) sulfide and excess hot $6\,M$ HCl.

Step E

22. Arsenic(III) sulfide and $6\,M$ aqueous NH_3 plus 6% H_2O_2.
23. Ammonium arsenate and magnesia mixture.

Step F

24. Tin(IV) chloride in dilute HCl reacts with iron.
25. Tin(II) chloride in dilute HCl and excess mercury(II) chloride in dilute HCl.
26. Antimony(III) chloride in dilute HCl reacts with iron.
27. Antimony and nitric acid plus oxalic acid.
28. A solution of trisoxalatoantimonic(III) acid and hydrogen sulfide.
29. Tin(IV) chloride in dilute HCl and excess oxalic acid.
30. Antimony(III) chloride in dilute HCl and excess oxalic acid.

Other Questions and Problems

31. A Group II unknown gave a yellow sulfide precipitate. Which cation(s) could be present?
32. A Group II unknown gave an orange sulfide precipitate. Which cation(s) could be present?

Analysis of Cation Group III

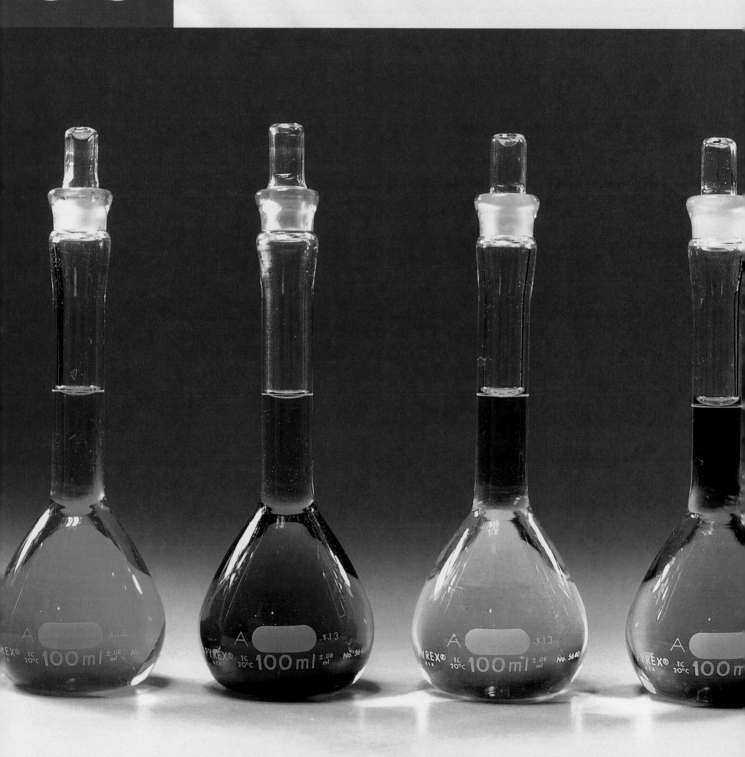

Aqueous solutions of some compounds that contain chromium. Left to right: chromium(II) chloride ($CrCl_2$) is blue; chromium(III) chloride ($CrCl_3$) is green; potassium chromate (K_2CrO_4) is yellow; potassium dichromate ($K_2Cr_2O_7$) is orange.

33-1 COMMON OXIDATION STATES OF THE METALS IN ANALYTICAL GROUP III

Cobalt exists in aqueous solutions primarily as cobalt(II) compounds. Cobalt(III) is such a strong oxidizing agent that it oxidizes water.

$$Co^{3+}(aq) + e^- \longrightarrow Co^{2+}(aq) \qquad E^0 = +1.82 \text{ V}$$

Only complex compounds containing Co(III) can exist in aqueous solution. In an alternative test for Co, the stable complex ion $[Co(NO_2)_6]^{3-}$ is formed. The hexaamminecobalt(III) ion, $[Co(NH_3)_6]^{3+}$, is stable in aqueous solutions.

Nickel compounds exist in aqueous solution primarily in the +2 oxidation state, although nickel(IV) oxide is precipitated from strongly basic solutions of strong oxidizing agents.

$$NiO_2(s) + 2H_2O + 2e^- \longrightarrow Ni(OH)_2(s) + 2OH^-(aq) \qquad E^0 = +0.49 \text{ V}$$

Nickel(IV) oxide is a powerful oxidizing agent in acidic solutions.

$$NiO_2(s) + 4H^+(aq) + 2e^- \longrightarrow Ni^{2+}(aq) + 2H_2O \qquad E^0 = +1.7 \text{ V}$$

The common oxidation states of iron are iron(II) and iron(III). In the +2 oxidation state, iron is a weak reducing agent. It is a fairly strong oxidizing agent in the +3 oxidation state.

$$Fe^{3+}(aq) + e^- \longrightarrow Fe^{2+}(aq) \qquad E^0 = +0.771 \text{ V}$$

Manganese exhibits a variety of oxidation states in aqueous solutions: +2, +3, +4, +6, and +7. The +2 oxidation state is the most stable. In its higher oxidation states, manganese is a strong oxidizing agent in acidic solutions.

$$MnO_2(s) + 4H^+(aq) + 2e^- \longrightarrow Mn^{2+}(aq) + 2H_2O \qquad E^0 = +1.23 \text{ V}$$
$$MnO_4^-(aq) + 8H^+(aq) + 5e^- \longrightarrow Mn^{2+}(aq) + 4H_2O \qquad E^0 = +1.51 \text{ V}$$

Higher oxidation state compounds of manganese are weaker oxidizing agents in basic solutions.

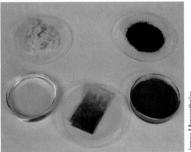

Solid $MnCl_2 \cdot 4H_2O$ and a concentrated solution of $MnCl_2$ (*left*). Solid $KMnO_4$ and a dilute solution of $KMnO_4$ (*right*). The cellulose in a piece of paper towel has reduced $KMnO_4$ to potassium manganate, K_2MnO_4, which is green (*center*). Under the photographer's hot lights, some K_2MnO_4 has been reduced to MnO_2, which is brown.

$$MnO_2(s) + 2H_2O + 2e^- \longrightarrow Mn(OH)_2(s) + 2OH^-(aq) \qquad E^0 = -0.05 \text{ V}$$

$$MnO_4^-(aq) + 2H_2O + 3e^- \longrightarrow MnO_2(s) + 4OH^-(aq) \qquad E^0 = +0.588 \text{ V}$$

Chromium also exhibits several oxidation states in aqueous solution (+2, +3, and +6). Chromium(II) is such a strong reducing agent that most Cr(II) compounds cannot exist in contact with atmospheric oxygen and water.

$$Cr^{3+}(aq) + e^- \longrightarrow Cr^{2+}(aq) \qquad E^0 = -0.41 \text{ V}$$

By contrast, chromium(VI) is an oxidizing agent in acidic solutions.

$$Cr_2O_7^{2-}(aq) + 14H^+(aq) + 6e^- \longrightarrow 2Cr^{3+}(aq) + 7H_2O \qquad E^0 = 1.33 \text{ V}$$

The +3 oxidation state is the most stable oxidation state of chromium in aqueous solutions.

Aluminum and zinc each exhibit only one oxidation state in their compounds in aqueous solutions: +3 for aluminum and +2 for zinc.

The procedures outlined in this chapter may be applied to a Group III known or unknown or to the centrifugate from the Groups I and II sulfide precipitation. The Group III flow chart is given in Figure 33-1.

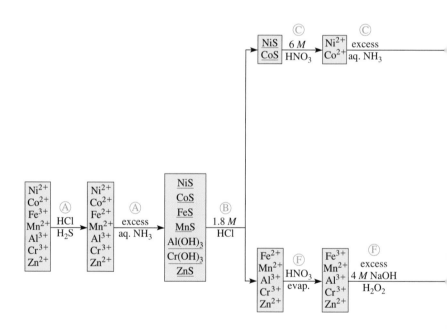

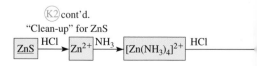

Figure 33-1 Analytical Group III flow chart. Circled letters refer to analytical steps. Species formed in confirmatory tests are shown with blue backgrounds. Precipitated species are underlined.

33-2 PRECIPITATION OF THE ANALYTICAL GROUP III CATIONS

The Group III cations are nickel, Ni^{2+}; cobalt, Co^{2+}; manganese, Mn^{2+}; iron, Fe^{2+} or Fe^{3+}; aluminum, Al^{3+}; chromium, Cr^{3+}; and zinc, Zn^{2+}. The concentration of sulfide ions in a solution of hydrogen sulfide that is also 0.3 M in hydrogen ions (Groups I and II separations) is too low to precipitate the Group III metal sulfides. However, in a buffered solution of ammonium sulfide, the sulfides of Ni^{2+}, Co^{2+}, Mn^{2+}, Fe^{2+}, and Zn^{2+} do precipitate, whereas Al^{3+} and Cr^{3+} precipitate as hydroxides. The Group III cations are known as the **insoluble basic sulfide group.** Table 33-1 shows solubility products for the freshly precipitated sulfides and hydroxides of Group III.

Zinc sulfide is the least soluble of the freshly precipitated Group III sulfides. We demonstrated that zinc sulfide does not precipitate from solutions that are saturated with H_2S and are 0.30 M in H^+ (see Example 31-1). However, when hydrogen sulfide is added to a solution that contains an excess of aqueous ammonia buffered with ammonium nitrate or ammonium chloride, a much higher concentration of sulfide ions is produced.

$$2NH_3(aq) + H_2S(aq) \longrightarrow 2NH_4^+(aq) + S^{2-}(aq)$$

TABLE 33-1	Solubility Products for Group III Sulfides and Hydroxides
Group III Compound	K_{sp}
CoS (α)	5.9×10^{-21}
NiS (α)	3.0×10^{-21}
FeS	4.9×10^{-18}
MnS	5.1×10^{-15}
Al(OH)$_3$	1.9×10^{-33}
Cr(OH)$_3$	6.7×10^{-31}
ZnS	1.1×10^{-21}

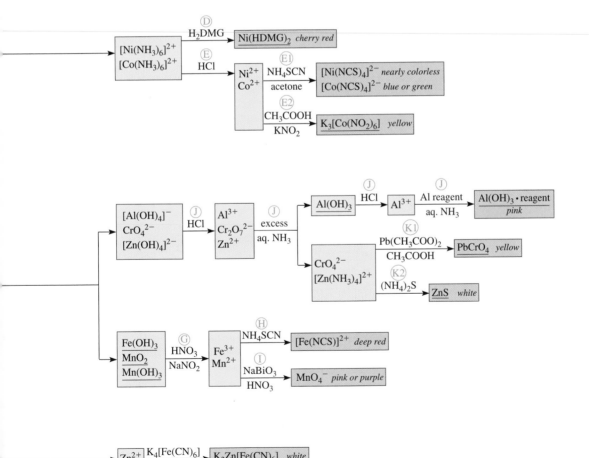

The sulfide ion concentration is sufficiently large that the solubility products for the sulfides of cobalt(II), nickel(II), manganese(II), iron(II), and zinc(II) are exceeded. So those sulfides precipitate from the solution. Similarly, the hydroxide ion concentration of the buffered solution of aqueous ammonia is sufficiently large that the hydroxides of aluminum and chromium(III) precipitate.

The sulfides of aluminum and chromium(III) cannot exist in contact with water because they hydrolyze completely and are converted to the corresponding hydroxides. Aluminum sulfide, Al_2S_3, and chromium(III) sulfide, Cr_2S_3, contain small, highly charged cations that hydrolyze extensively (Section 18-11 in the textbook) as well as an anion that hydrolyzes extensively.

Heating thioacetamide in the acidic solution forms H_2S. In acidic solution the H_2S reduces iron(III) ions to iron(II) ions.

$$2Fe^{3+}(aq) + H_2S(aq) \longrightarrow 2Fe^{2+}(aq) + S(s) + 2H^+(aq)$$

The addition of aqueous ammonia to the acidic solution of hydrogen sulfide produces a high concentration of sulfide ions, and the Group III precipitation reactions occur.[1]

$$Ni^{2+} + S^{2-} \longrightarrow NiS(s) \quad \text{black}$$
$$Co^{2+} + S^{2-} \longrightarrow CoS(s) \quad \text{black}$$
$$Fe^{2+} + S^{2-} \longrightarrow FeS(s) \quad \text{black}$$
$$Mn^{2+} + S^{2-} \longrightarrow MnS(s) \quad \text{salmon}$$
$$Zn^{2+} + S^{2-} \longrightarrow ZnS(s) \quad \text{white}$$
$$Al^{3+} + 3NH_3 + 3H_2O \longrightarrow Al(OH)_3(s) + 3NH_4^+$$
$$\text{white}$$
$$Cr^{3+} + 3NH_3 + 3H_2O \longrightarrow Cr(OH)_3(s) + 3NH_4^+$$
$$\text{gray-green}$$

Although magnesium hydroxide, $Mg(OH)_2$ is an insoluble hydroxide, it is not precipitated in Group III because the solution is strongly buffered with ammonium chloride and ammonium sulfide. The high concentration of ammonium ions from these salts inhibits the ionization of aqueous ammonia so that the solubility product for magnesium hydroxide is not exceeded. Consequently, magnesium ions are found in Group V rather than in Group III.

Several of the cations in Group III are derived from transition metals and are colored in aqueous solutions. Their colors may give valuable clues about their presence or absence in unknowns. The colors commonly associated with these hydrated ions are as follows.

[1]*Alert students may wonder about the reaction of the Group III cations with aqueous ammonia in the buffered solution (Sections 19-2 and 25-2). Aqueous NH_3 is added to an acidic solution that contains the Group III cations and is saturated with H_2S, an acid. No doubt, other reactions (similar to those illustrated for Ni^{2+}) occur to some extent.*

$$Ni^{2+}(aq) + 2NH_3(aq) + 2H_2O \longrightarrow Ni(OH)_2(s) + 2NH_4^+(aq)$$
$$Ni(OH)_2(s) + 6NH_3(aq) \longrightarrow [Ni(NH_3)_6]^{2+}(aq) + 2OH^-(aq)$$
$$[Ni(NH_3)_6]^{2+}(aq) + S^{2-}(aq) \longrightarrow NiS(s) + 6NH_3(aq)$$

However, the net reaction may be represented as

$$Ni^{2+}(aq) + S^{2-}(aq) \longrightarrow NiS(s)$$

We have chosen to represent the Group III precipitation reactions in the simplest way possible.

Throughout this supplement, references to chapters or sections prior to Chapter 29 refer to *General Chemistry* by Whitten, Davis, Peck, and Stanley.

Co^{2+}	red-pink	Mn^{2+}	pale pink
Ni^{2+}	green	Cr^{3+}	deep blue (NO_3^- solutions)
Fe^{2+}	pale green		deep green (Cl^- solutions)
Fe^{3+}	orchid (NO_3^- solutions)		
	yellow (Cl^- solutions)		

The Group III known and unknown solutions are usually prepared by dissolving nitrates of Group III metals in dilute nitric acid solution to suppress hydrolysis of the cations. General unknowns may contain either chlorides or nitrates.

Solutions containing Al^{3+} and Zn^{2+} ions are colorless unless the anions that occur with them impart color. Chromium(III) ion is so intensely colored that its presence usually masks the presence of other colored ions in Group III.

Step A

The solution may contain one or more of Co^{2+}, Ni^{2+}, Fe^{3+}, Mn^{2+}, Al^{3+}, Cr^{3+}, and Zn^{2+}.

(1) If the analysis is to be performed on the centrifugate from the Groups I and II sulfide precipitation, add 8 drops of 5% thioacetamide solution and heat the mixture in a water bath for 3 min.

(2) If the analysis is to be performed on a known or unknown solution containing only Group III cations, add two drops of 6 M HCl to 10 drops of the solution, dilute to 1 mL, and add 8 drops of 5% thioacetamide solution. Heat the solution in a hot water bath for 3 min.

Add 5 drops of 15 M aqueous ammonia to the solution produced by either treatment (1) or (2), and stir vigorously. Test the solution to determine that it is basic to litmus. If not, continue the dropwise addition of 15 M aqueous ammonia until it is basic. Heat the mixture in a water bath for 5 min, stirring frequently. Separate* the precipitate, and wash it with a mixture of 10 drops of water and 1 drop of 3 M aqueous ammonia. Save the solution in which the precipitation occurred for the analysis of Groups IV and V if the unknown is a general unknown.[†] Otherwise it may be discarded. The precipitate is treated according to Step B.

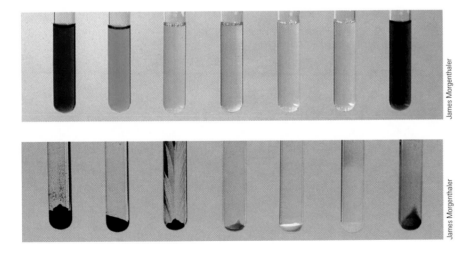

Solutions of Group III nitrates (*top*). *Left to right:* Co^{2+}, Ni^{2+}, Fe^{3+}, Mn^{2+}, Al^{3+}, Zn^{2+}, Cr^{3+}. The Group III precipitates (*bottom*). *Left to right:* CoS (*black*), NiS (*black*), FeS (*black*), MnS (*salmon*), ZnS (*white*), $Al(OH)_3$ (*white*), $Cr(OH)_3$ (*gray-green*).

*Both cobalt(II) sulfide and nickel(II) sulfide tend to form colloidal suspensions. A colloidal suspension of either or both of these sulfides is usually intensely blue. If a colloidal suspension develops, the mixture should be heated for several minutes until it becomes clear (transparent) before it is centrifuged and separated.

[†]Sulfide ions can be oxidized to sulfate ions by atmospheric O_2 in basic solution ($S^{2-} + 2O_2 \longrightarrow SO_4^{2-}$). Sulfate ions react with barium ions to form insoluble barium sulfate and cause Ba^{2+} ions to be missed in Group IV. Acidify the solution that contains Groups IV–V with 6 M HCl, and boil it until all the H_2S has been expelled.

33-3 SEPARATION OF COBALT AND NICKEL

Hydrogen sulfide precipitates the sulfides of cobalt and nickel completely *only* from basic solutions. However, these sulfides are only very slightly soluble in dilute HCl. Although they precipitate in forms that are soluble in weakly acidic solutions, they change rapidly into other crystalline forms that are much less soluble. For example, the solubility product for NiS (α) is 3.0×10^{-21}, that for NiS (β) is 1.0×10^{-26}, and that for NiS (γ) is 2.0×10^{-28}. The fact that CoS and NiS are rapidly converted to much less soluble crystalline forms is the basis for separating them from the other Group III cations. The sulfides of Fe^{2+}, Mn^{2+}, and Zn^{2+} and the hydroxides of Al^{3+} and Cr^{3+} are soluble in 1.8 M HCl; the sulfides of Co^{2+} and Ni^{2+} are not.

$$MnS(s) + 2H^+ \longrightarrow Mn^{2+} + H_2S(g)$$
$$FeS(s) + 2H^+ \longrightarrow Fe^{2+} + H_2S(g)$$
$$Al(OH)_3(s) + 3H^+ \longrightarrow Al^{3+} + 3H_2O$$
$$Cr(OH)_3(s) + 3H^+ \longrightarrow Cr^{3+} + 3H_2O$$
$$ZnS(s) + 2H^+ \longrightarrow Zn^{2+} + H_2S(g)$$

These dissolution reactions involve the formation of covalent compounds, H_2S and H_2O. H_2S is volatile and escapes from the solution as it is formed. This drives the reactions for the dissolution of these sulfides far to the right.

About 1 drop of water remains in the precipitate from Step A, so the HCl concentration is $(3/10) \times 6$ M = 1.8 M.

Step B
Use the precipitate from Step A, containing CoS, NiS, FeS, MnS, Al(OH)$_3$, Cr(OH)$_3$, and ZnS. Add 6 drops of water *followed by* 3 drops of 6 M HCl to the precipitate, and stir vigorously for 2 min. Separate the mixture immediately and save the centrifugate for Step F. Add 6 drops of water *followed by* 3 drops of 6 M HCl to the residue, and stir it vigorously for 1 min. Separate immediately, and add the centrifugate to that obtained in the first extraction. The precipitate is treated according to Step C, and the combined extracts are used in Step F.

33-4 IDENTIFICATION OF COBALT AND NICKEL

Cobalt(II) and nickel(II) sulfides dissolve in hot nitric acid, which oxidizes the sulfide ions to free sulfur.

$$3CoS(s) + 8H^+ + 2NO_3^- \longrightarrow 3Co^{2+} + 3S(s) + 2NO + 4H_2O$$
$$3NiS(s) + 8H^+ + 2NO_3^- \longrightarrow 3Ni^{2+} + 3S(s) + 2NO + 4H_2O$$

These reactions are similar to the reactions by which the Group I sulfides were dissolved. After the sulfides of cobalt and nickel are dissolved, the solution is boiled for several minutes to remove the oxides of nitrogen. An excess of aqueous ammonia is then added to convert cobalt(II) and nickel(II) ions to hexaammine complexes (Section 25-2 in the textbook).

The reactions of cobalt(II) ions are identical to these.

$$Ni^{2+} + 2NH_3 + 2H_2O \longrightarrow Ni(OH)_2(s) + 2NH_4^+$$
$$\text{lim. amt.}$$
$$Ni(OH)_2(s) + 6NH_3 \longrightarrow [Ni(NH_3)_6]^{2+} + 2OH^-$$
$$\text{excess}$$

Dimethylglyoxime is an organic compound that reacts with nickel(II) ions in aqueous ammonia to form an insoluble, bright cherry red complex compound. The hydrogen atoms in oximes

$$\underset{/}{\overset{\backslash}{C}}=N-O-H$$

are very weakly acidic. Dimethylglyoxime may be abbreviated H_2DMG.

$$2\ \begin{matrix} CH_3-C=N-O-H \\ | \\ CH_3-C=N-O-H \end{matrix} + [Ni(NH_3)_6]^{2+} \longrightarrow$$

(reaction product structure)

$Ni(HDMG)_2(s)$, bright pink-red

The arrows indicate coordinate covalent bonds, and the dashed lines indicate hydrogen bonds.

Cobalt (II) ions form a brown complex with dimethylglyoxime, but the complex does not interfere with the test for nickel(II) ions.

The simplest confirmatory test for cobalt(II) ions involves their reaction with thiocyanate ions to form tetrathiocyanatocobaltate(II) ions, $[Co(NCS)_4]^{2-}$. The latter are blue or blue-green in 50% H_2O/50% acetone solution.

$$Co^{2+} + 4SCN^- \longrightarrow [Co(NCS)_4]^{2-} \qquad \text{blue or blue-green}$$

If the separation of cobalt(II) and nickel(II) ions from the other Group III cations is less than complete, iron(III) ions interfere with the test for cobalt by forming bright red complex ions, $[Fe(NCS)]^{2+}$.

$$Fe^{3+} + SCN^- \rightleftharpoons [Fe(NCS)]^{2+} \qquad \text{red}$$

This interference may be eliminated by the addition of fluoride ions, which convert the red complex ions into very stable, colorless hexafluoroferrate(III) ions, $[FeF_6]^{3-}$.

$$\underset{\text{red}}{[Fe(NCS)]^{2+}} + 6F^- \rightleftharpoons \underset{\text{colorless}}{[FeF_6]^{3-}} + SCN^-$$

We have written the formula for the thiocyanate ion as SCN^-, which is the accepted way to write it. However, complex species containing SCN^- ions are usually written $[Co(NCS)_4]^{2-}$ to emphasize that the nitrogen atom is the donor atom in most complexes of this kind.

Once the interference by the red $[Fe(NCS)]^{2+}$ ions has been eliminated, the characteristic blue color of the $[Co(NCS)_4]^{2-}$ ions is observed easily.

An alternative test for cobalt(II) ions involves the formation of a golden yellow precipitate, $K_3[Co(NO_2)_6]$, in mildly acidic solution. Cobalt(II) ions form hexanitrocobaltate(II) ions, $[Co(NO_2)_6]^{4-}$, with excess nitrite ions.

$$Co^{2+} + 6NO_2^- \longrightarrow [Co(NO_2)_6]^{4-}$$

These are oxidized to hexanitrocobaltate(III) ions, $[Co(NO_2)_6]^{3-}$, by excess NO_2^-.

$$[Co(NO_2)_6]^{4-} + NO_2^- + 2H^+ \longrightarrow [Co(NO_2)_6]^{3-} + NO + H_2O$$

These complex ions are "nitro" complexes, which tells us that the nitrogen atoms are coordinated to cobalt. If coordination were through oxygen atoms, the complex ions would be called "nitrito" complexes.

These in turn react with potassium ions in the buffered acetic acid solution to form one of a very few insoluble potassium compounds, potassium hexanitrocobaltate(III).

$$3K^+ + [Co(NO_2)_6]^{3-} \longrightarrow K_3[Co(NO_2)_6](s) \qquad \text{golden yellow}$$

The confirmatory test for nickel, Ni(HDMG)$_2$.

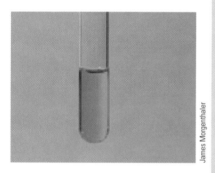

The confirmatory test for cobalt, [Co(NCS)$_4$]$^{2-}$.

Step C

Use the residue from Step B, containing CoS and NiS. Add 8 drops of 6 M HNO$_3$ to the residue, stir thoroughly, and heat the mixture in a water bath until the black sulfide precipitate has dissolved completely. Stir the solution regularly while it is heated for an additional 3 min. Add 10 drops of water and just enough 3 M aqueous ammonia to make the solution basic to litmus. Avoid a large excess of aqueous ammonia. Centrifuge the solution and discard any sulfur that remains.

Step D

Test for nickel. Place 6 drops of the solution from Step C in a test tube and add 3 drops of dimethylglyoxime. The formation of a cherry red precipitate confirms the presence of nickel(II) ions. If cobalt is present and nickel is absent, the solution will turn brown when dimethylglyoxime is added. Don't confuse a brown solution with the bright cherry red precipitate obtained when nickel is present.

Step E

Test for cobalt. Perform *either* test (1) or (2) as directed by your instructor.

(1) Carefully acidify the remainder of the solution from Step C, which may contain cobalt(II) and nickel(II) ions, with 6 M HCl (avoid excess HCl). Then add 5 drops of 4 M NH$_4$SCN solution. Double the volume of solution by adding acetone. Place a clean finger over the top of the test tube and invert it. If cobalt(II) ions are present, the solution will turn blue or blue-green, depending on its acidity. (If the solution is red owing to the incomplete removal of iron, add 4 M KF solution dropwise [mix thoroughly after each drop is added] until the red color disappears.)

(2) Transfer the remainder of the solution to a small beaker or evaporating dish, and carefully evaporate it to approximately 1 drop, *but not to dryness*. Be prepared to add a few drops of water to prevent the complete evaporation of liquid. Allow the mixture to cool. Add 3 drops of water, 4 drops of 2 M KCH$_3$COO, and enough 6 M CH$_3$COOH to make the solution acidic to litmus. Mix well, transfer the liquid to a clean test tube, and centrifuge if necessary to remove any solid. Double the volume by adding 6 M KNO$_2$, mix well, and place the solution in the water bath for 3 min. Remove the test tube from the water bath, and allow it to stand for 15 min. A golden yellow precipitate of K$_3$[Co(NO$_2$)$_6$], often slow in forming, confirms the presence of cobalt.

33-5 SEPARATION OF IRON AND MANGANESE FROM ALUMINUM, CHROMIUM, AND ZINC

The solution from Step B that contains Fe^{2+}, Mn^{2+}, Al^{3+}, Cr^{3+}, and Zn^{2+} ions also contains H$_2$S. H$_2$S must be removed before iron and manganese are separated from aluminum, chromium, and zinc. This is accomplished by evaporating the solution with nitric acid, which oxidizes H$_2$S to free sulfur and Fe^{2+} ions to Fe^{3+} ions.

$$3H_2S + 2H^+ + 2NO_3{}^- \longrightarrow 3S(s) + 2NO + 4H_2O$$

$$3Fe^{2+} + 4H^+ + NO_3{}^- \longrightarrow 3Fe^{3+} + NO + 2H_2O$$

The residue is dissolved in water and treated with 4 M NaOH, which results initially in the precipitation of hydroxides of the five metal ions.

$$Fe^{3+} + 3OH^- \longrightarrow Fe(OH)_3(s) \qquad \text{red-brown}$$
$$Mn^{2+} + 2OH^- \longrightarrow Mn(OH)_2(s) \qquad \text{white}$$
$$Al^{3+} + 3OH^- \longrightarrow Al(OH)_3(s) \qquad \text{white}$$
$$Cr^{3+} + 3OH^- \longrightarrow Cr(OH)_3(s) \qquad \text{gray-green}$$
$$Zn^{2+} + 2OH^- \longrightarrow Zn(OH)_2(s) \qquad \text{white}$$

The continued addition of $4\,M$ NaOH results in the dissolution of the three *amphoteric* hydroxides (Section 10-6 in the textbook).

$$Al(OH)_3(s) + OH^- \longrightarrow [Al(OH)_4]^-$$
$$Cr(OH)_3(s) + OH^- \longrightarrow [Cr(OH)_4]^-$$
$$Zn(OH)_2(s) + 2OH^- \longrightarrow [Zn(OH)_4]^{2-}$$

The strongly basic mixture is then treated with hydrogen peroxide, a strong oxidizing agent in basic solution. The tetrahydroxochromate(III) ions are oxidized to chromate ions (yellow).

$$2[Cr(OH)_4]^- + 3H_2O_2 + 2OH^- \longrightarrow 2CrO_4^{2-} + 8H_2O$$

Manganese(II) hydroxide is oxidized to a mixture of manganese(III) hydroxide (black) and manganese(IV) oxide (dark brown), commonly called manganese dioxide. Both of these compounds are quite insoluble.

$$2Mn(OH)_2(s) + H_2O_2 \longrightarrow 2Mn(OH)_3(s)$$
$$Mn(OH)_2(s) + H_2O_2 \longrightarrow MnO_2(s) + 2H_2O$$

Step F

Use the solution from Step B, containing Mn^{2+}, Fe^{2+}, Al^{3+}, Cr^{3+}, and Zn^{2+}. Add 5 drops of $16\,M$ HNO_3 to the solution and evaporate to a moist residue. Cool, add 10 drops of water, mix well, and then add $4\,M$ NaOH dropwise, with constant stirring, until a precipitate forms. Then add another 10 drops of $4\,M$ NaOH. Stir thoroughly, and add 10 drops of 6% hydrogen peroxide to the mixture. Heat it in a water bath for 5 min, stirring frequently. Separate the mixture and save the centrifugate for Step J. Wash the residue with 1 drop of $4\,M$ NaOH in 10 drops of water. Treat the precipitate according to Step G.

33-6   IDENTIFICATION OF MANGANESE AND IRON

The residue from Step F contains a mixture of MnO_2, $Mn(OH)_3$, and $Fe(OH)_3$. It is treated with a mixture of HNO_3 and $NaNO_2$. The iron(III) hydroxide dissolves in nitric acid in a typical acid–base reaction.

$$Fe(OH)_3(s) + 3H^+ \longrightarrow Fe^{3+} + 3H_2O$$

In acidic solution nitrite ions reduce both manganese(III) and manganese(IV) to manganese(II). Nitrite ions are oxidized to nitrate ions.

$$2Mn(OH)_3(s) + 4H^+ + NO_2^- \longrightarrow 2Mn^{2+} + NO_3^- + 5H_2O$$
$$MnO_2(s) + 2H^+ + NO_2^- \longrightarrow Mn^{2+} + NO_3^- + H_2O$$

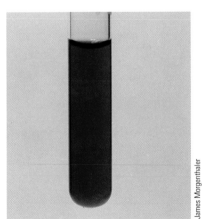

The confirmatory test for iron, $[Fe(NCS)]^{2+}$.

James Morgenthaler

The confirmatory test for Fe^{3+} is its reaction with thiocyanate ions, SCN^-, to produce deep red complex ions, $[Fe(NCS)]^{2+}$. (This was discussed as an interference in the test for cobalt in Section 33-4.) The presence of Mn^{2+} ions does not interfere with the test for Fe^{3+} ions, and Fe^{3+} ions do not interfere with the test for Mn^{2+} ions. Traces of iron contaminate almost everything, and a light pink color may occur even when iron is not present in the unknown. If a light pink solution is obtained in the test for iron(III) ions, a blank test should be run on the reagents, using everything except the solution that contains Fe^{3+} ions, to determine whether the pink color is due to impurities in the reagents.

Manganese is detected by converting it to permanganate ions, MnO_4^-, which impart a purple color to the solution. In very dilute solutions permanganate ions appear pink. A powerful oxidizing agent is required to oxidize Mn^{2+} to MnO_4^- ions. Sodium bismuthate, $NaBiO_3$, is commonly used.

$$NaBiO_3(s) + 6H^+(aq) + 2e^- \longrightarrow Bi^{3+}(aq) + Na^+(aq) + 3H_2O \qquad E^0 \approx +1.61 \text{ V}$$

$$2Mn^{2+} + 5BiO_3^- + 14H^+ \longrightarrow 2MnO_4^- + 5Bi^{3+} + 7H_2O$$

The bismuthate ion, BiO_3^-, is a very powerful oxidizing agent, one of the few that will oxidize Mn^{2+} ions to MnO_4^- ions in acidic solution; the permanganate ion itself is a very strong oxidizing agent.

The confirmatory test for manganese, MnO_4^-.

James Morgenthaler

Step G

Use the residue from Step F, containing $Fe(OH)_3$, $Mn(OH)_3$, and MnO_2. Add 8 drops of 6 M HNO_3 and 2 drops of 1 M $NaNO_2$ to the residue. Heat the mixture in a water bath for 6 min, stirring frequently. Cool the test tube under running water. Add 10 drops of H_2O. Stir.

Step H

Test for iron. Transfer 10 drops of the solution from Step G to a clean test tube and add 3 drops of 4 M NH_4SCN. A dark red color develops in the solution if iron is present in the unknown. A faint pink indicates a trace of iron, probably due to impurities, and should be ignored. If a deep red color develops and then fades rapidly, this indicates the presence of unreacted nitrite ions; but the appearance of the deep red color (even briefly) is taken as confirmation of the presence of iron.

Step I

Test for manganese. To the remainder of the solution from Step G add 2 drops of 6 M HNO_3 and some solid $NaBiO_3$. Sufficient $NaBiO_3$ should be added that the solution is muddy in appearance after it has been stirred thoroughly. Centrifuge the solution and observe the color above the dark brown $NaBiO_3$. The solution should be pink to purple, depending on the amount of manganese present in the unknown.

33-7 SEPARATION AND IDENTIFICATION OF ALUMINUM

Hydrochloric acid is added to the basic solution from the H_2O_2 oxidation (Step F) to neutralize the excess sodium hydroxide. It also destroys the hydroxocomplexes of Al and Zn, converting them first to the insoluble hydroxides

$$[Al(OH)_4]^- + H^+ \longrightarrow Al(OH)_3(s) + H_2O$$

$$[Zn(OH)_4]^{2-} + 2H^+ \longrightarrow Zn(OH)_2(s) + 2H_2O$$

These insoluble hydroxides then dissolve as an excess of HCl is added.

$$Al(OH)_3(s) + 3H^+ \longrightarrow Al^{3+} + 3H_2O$$

$$Zn(OH)_2(s) + 2H^+ \longrightarrow Zn^{2+} + 2H_2O$$

Chromate ions, CrO_4^{2-}, are converted to dichromate ions, $Cr_2O_7^{2-}$, by acids.

$$2CrO_4^{2-} + 2H^+ \rightleftharpoons Cr_2O_7^{2-} + H_2O$$

yellow   orange

The addition of aqueous ammonia reprecipitates aluminum hydroxide. It converts $Cr_2O_7^{2-}$ ions back to CrO_4^{2-} (reverses the previous reaction) and forms an ammine complex with zinc ions (Section 25-2 in the textbook).

$$Al^{3+} + 3NH_3 + 3H_2O \longrightarrow Al(OH)_3(s) + 3NH_4^+$$

$$Zn^{2+} + 2NH_3 + 2H_2O \longrightarrow Zn(OH)_2(s) + 2NH_4^+$$

$$Zn(OH)_2(s) + 4NH_3 \longrightarrow [Zn(NH_3)_4]^{2+} + 2OH^-$$

$$Cr_2O_7^{2-} + 2NH_3 + H_2O \rightleftharpoons 2CrO_4^{2-} + 2NH_4^+$$

It is especially important that an excess of aqueous ammonia be added so that all of the zinc hydroxide dissolves and none is left behind with aluminum hydroxide.

In the hydrogen peroxide oxidation (Step F), a strongly basic solution was boiled for several minutes. Strong bases dissolve glass to a slight extent to form soluble silicates. Acidification of solutions that contain silicates produces hydrated silicon dioxide, $SiO_2 \cdot xH_2O$, a nearly colorless, gelatinous substance that looks very much like aluminum hydroxide. After the solution has been acidified with HCl, it is mixed thoroughly and centrifuged for a few minutes. Any precipitate of silicon dioxide should be discarded.

To test for aluminum, we first dissolve its hydroxide in hydrochloric acid.

$$Al(OH)_3(s) + 3H^+ \longrightarrow Al^{3+} + 3H_2O$$

Then aluminum hydroxide is precipitated by the addition of aqueous ammonia in the presence of **aluminum reagent,** a red dye that is physically trapped by the gelatinous aluminum hydroxide.

$$Al^{3+} + 3NH_3 + 3H_2O \longrightarrow Al(OH)_3(s) + 3NH_4^+$$

Step J

Use the solution from Step F, containing $[Al(OH)_4]^-$, CrO_4^{2-}, and $[Zn(OH)_4]^{2-}$. Add enough 6 M HCl to the solution to make it acidic to litmus, and then add an extra 2 drops. Stir vigorously for 1 min. Centrifuge the solution and observe carefully to determine whether enough silicon dioxide, SiO_2, has precipitated that it can be separated from the solution. If so, withdraw the solution and discard the silicon dioxide. Add 6 M aqueous ammonia to the solution until it is basic to litmus, then add 8 drops in excess. Stir vigorously for 1 min. The formation of a white (more accurately, nearly colorless), gelatinous precipitate is likely due to aluminum hydroxide. Separate the mixture and keep the centrifugate for Step K.

To test for aluminum, dissolve the precipitate in 2 drops of 6 M HCl. Add 5 drops of water and 2 drops of aluminum reagent, followed by enough 6 M aqueous ammonia to make the solution basic. The formation of an orange to red precipitate confirms the presence of aluminum.

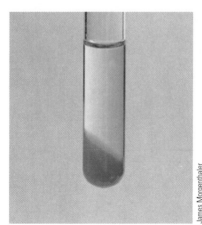

James Morgenthaler

The confirmatory test for aluminum.

33-8 IDENTIFICATION OF CHROMIUM AND ZINC

The solution that contains zinc and chromium also contains an excess of aqueous ammonia. This must be neutralized before the solution can be tested for chromate ions. (Zinc ions do not interfere with the test.) We test for chromate ions by adding lead acetate, a covalent compound.

$$Pb(CH_3COO)_2 + CrO_4{}^{2-} \longrightarrow PbCrO_4(s) + 2CH_3COO^-$$

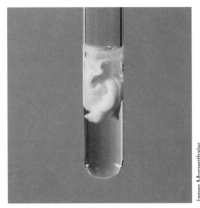

Zinc ions are precipitated from the strongly ammoniacal solution by ammonium sulfide, $(NH_4)_2S$. Ammonium sulfide is susceptible to air oxidation. The reagent should be yellow, and there should be no flakes of sulfur floating in it. If there is any question about the reagent, test it by adding 2 drops of the known zinc solution to 10 drops of water and then adding 2 drops of $(NH_4)_2S$. The formation of a large amount of white precipitate, ZnS, indicates that the reagent is all right.

$$[Zn(NH_3)_4]^{2+} + S^{2-} \longrightarrow ZnS(s) + 4NH_3$$

Traces of the Group III ions that form black sulfides—Fe^{3+}, Ni^{2+}, and Co^{2+}—interfere with the test for zinc. They make it impossible to see the white precipitate of zinc sulfide. If the precipitate that should be zinc sulfide is dark, it is then treated with 1.8 M HCl. The clear, colorless solution can be separated from any dark precipitate. The solution is then boiled to expel hydrogen sulfide as completely as possible, the excess HCl is neutralized with aqueous ammonia, and finally the solution is again made slightly acidic with HCl. The addition of potassium hexacyanoferrate(II), $K_4[Fe(CN)_6]$, results in the formation of a white precipitate, $K_2Zn[Fe(CN)_6]$, which proves the presence of zinc.

The confirmatory test for chromium, $PbCrO_4$.

$$Zn^{2+} + 2K^+ + [Fe(CN)_6]^{4-} \longrightarrow K_2Zn[Fe(CN)_6](s) \qquad \text{white}$$

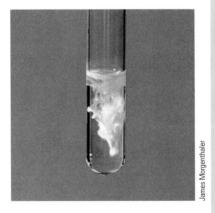

The confirmatory test for zinc, ZnS.

Step K

Use the solution from Step J, containing $CrO_4{}^{2-}$ and $[Zn(NH_3)_4]^{2+}$.

(1) *Test for chromium.* To half of the solution add 6 M CH_3COOH until the solution is acidic to litmus. Add 2 drops of 0.1 M $Pb(CH_3COO)_2$. The presence of chromium is verified by the appearance of a bright yellow precipitate of $PbCrO_4$.

(2) *Test for zinc.* To the other half of the solution add 5 drops of 5% $(NH_4)_2S$. The formation of a white precipitate, ZnS, that is soluble in 1.8 M HCl proves that zinc is present. If the precipitate is not white, separate it from the liquid and discard the liquid. Add 7 drops of water followed by 3 drops of 6 M HCl; stir the mixture thoroughly for 1 min; centrifuge; and withdraw the clear, colorless liquid. Discard any precipitate. Boil the liquid until the H_2S has been expelled and then add sufficient 3 M aqueous ammonia to make it basic to litmus. Now add 10 drops of 1 M HCl, check to determine that the solution is acidic to litmus, and add 5 drops of 0.2 M $K_4[Fe(CN)_6]$. If zinc is present, it precipitates as a white compound, $K_2Zn[Fe(CN)_6]$.

Exercises

General Questions on Analytical Group III

1. List the common oxidation states exhibited by the metals whose cations occur in Group III. Indicate those that are usually reducing oxidation states by (R), those that are considered "stable" oxidation states by (S), and those that are oxidizing by (O).

2. Known and unknown solutions contain 5 milligrams of each cation per milliliter of solution, except that of Al^{3+}, which contains 10 mg/mL. Calculate the molarity of each Group III cation in such solutions.

3. Which of the freshly precipitated Group III sulfides is (a) most soluble? (b) least soluble?

4. Why are Al^{3+} and Cr^{3+} ions precipitated in Group III as hydroxides rather than as sulfides?

5. List all Group III cations that (as NO_3^- or Cl^- salts) fit the following descriptions.
   (a) Colorless solutions.
   (b) Green solutions.
   (c) Soluble in excess aqueous NH_3.
   (d) Insoluble in excess aqueous NH_3.
   (e) Soluble in excess NaOH solution.
   (f) Insoluble in excess NaOH solution.
   (g) Soluble in both excess aqueous NH_3 and excess NaOH solution.
   (h) Insoluble in both excess aqueous NH_3 and excess NaOH solution.
   (i) Soluble in excess aqueous NH_3, but not in excess NaOH solution.
   (j) Soluble in excess NaOH solution, but not in excess aqueous NH_3.
   (k) Pink solution gives a black sulfide.
   (l) Green solution gives a black sulfide.
   (m) Oxidized by H_2O_2.
   (n) Reduced by H_2S.
   (o) Neither oxidized nor reduced in Group III procedures.

Analytical Group III Reactions

Some reactions occur at several points in the analysis scheme for Group III cations. Therefore, these reactions are arranged somewhat differently than earlier group reactions. Write balanced net ionic equations for the reactions that occur when the following substances are mixed in aqueous solutions. Indicate the colors of all precipitates and complex ions.

Reactions with a Limited Amount of NH₃(aq)

6. Cobalt(II) nitrate + limited amount of aqueous NH_3.
7. Nickel(II) nitrate + limited amount of aqueous NH_3.
8. Manganese(II) nitrate + limited amount of aqueous NH_3.
9. Iron(II) nitrate + limited amount of aqueous NH_3.
10. Aluminum nitrate + limited amount of aqueous NH_3.
11. Chromium(III) nitrate + limited amount of aqueous NH_3.
12. Zinc nitrate + limited amount of aqueous NH_3.

Hydroxides with Excess NH₃(aq)

13. Cobalt(II) hydroxide + excess aqueous NH_3.
14. Nickel(II) hydroxide + excess aqueous NH_3.
15. Zinc hydroxide + excess aqueous NH_3.

Hydroxides with Strong Acids

16. Cobalt(II) hydroxide + nitric acid.
17. Nickel(II) hydroxide + nitric acid.
18. Manganese(II) hydroxide + nitric acid.
19. Iron(III) hydroxide + nitric acid.
20. Aluminum hydroxide + nitric acid.
21. Chromium(III) hydroxide + nitric acid.
22. Zinc hydroxide + nitric acid.

Ammine Complexes with Strong Acids

23. Hexaamminecobalt(II) hydroxide + HCl.
24. Hexaamminenickel(II) hydroxide + HCl.
25. Tetraamminezinc hydroxide + HCl.

Reactions with a Limited Amount of NaOH

26. Cobalt(II) nitrate + limited amount of NaOH.
27. Nickel(II) nitrate + limited amount of NaOH.
28. Manganese(II) nitrate + limited amount of NaOH.
29. Iron(III) nitrate + limited amount of NaOH.
30. Aluminum nitrate + limited amount of NaOH.
31. Chromium(III) nitrate + limited amount of NaOH.
32. Zinc nitrate + limited amount of NaOH.

Hydroxides with Excess NaOH (Table 29-3)

33. Cobalt(II) hydroxide + excess NaOH.
34. Aluminum hydroxide + excess NaOH.
35. Chromium(III) hydroxide + excess NaOH.
36. Zinc hydroxide + excess NaOH.

Hydroxocomplexes with Strong Acids

37. Sodium tetrahydroxoaluminate + HCl.
38. Sodium tetrahydroxochromate(III) + HCl.
39. Sodium tetrahydroxozincate + HCl.

Other Reactions

40. Iron(III) nitrate + hydrogen sulfide in acidic solution.
41. Cobalt(II) nitrate + ammonium sulfide in buffered aqueous NH_3.
42. Nickel(II) nitrate + ammonium sulfide in buffered aqueous NH_3.

43. Iron(II) nitrate + ammonium sulfide in buffered aqueous NH_3.
44. Manganese(II) nitrate + ammonium sulfide in buffered aqueous NH_3.
45. Aluminum nitrate + ammonium sulfide in buffered aqueous NH_3.
46. Chromium(III) nitrate + ammonium sulfide in buffered aqueous NH_3.
47. Zinc nitrate + ammonium sulfide in buffered aqueous NH_3.
48. Iron(II) sulfide + 1.8 M HCl.
49. Manganese(II) sulfide + 1.8 M HCl.
50. Zinc sulfide + 1.8 M HCl.
51. Cobalt(II) sulfide + hot 6 M HNO_3.
52. Nickel(II) sulfide + hot 6 M HNO_3.
53. Cobalt(II) chloride + ammonium thiocyanate.
54. Cobalt(II) acetate + potassium nitrite (three reactions).
55. Hexaamminenickel(II) hydroxide + dimethylglyoxime (H_2DMG).
56. Iron(II) hydroxide + hydrogen peroxide in excess NaOH.
57. Manganese(II) hydroxide + hydrogen peroxide in excess NaOH (two reactions).
58. Sodium tetrahydroxochromate(III) + hydrogen peroxide in excess NaOH.
59. Manganese(III) hydroxide + sodium nitrite + HNO_3.
60. Manganese(IV) oxide + sodium nitrite + HNO_3.
61. Manganese(II) nitrate + sodium bismuthate + HNO_3.
62. Iron(III) nitrate + ammonium thiocyanate.
63. Sodium chromate + hydrochloric acid.
64. Sodium dichromate + aqueous NH_3.
65. Sodium chromate + lead(II) acetate.
66. Tetraamminezinc hydroxide + ammonium sulfide.
67. Zinc chloride + potassium hexacyanoferrate(II).

Other Questions and Problems

68. What is the color of each of the following? Aqueous solutions are indicated by (aq). $Ni(NO_3)_2$(aq); NiS;

$Ni(HDMG)_2$; $Co(NO_3)_2$(aq); CoS; $[Co(NCS)_4]^{2-}$(aq); $K_3[Co(NO_2)_6]$; $Fe(NO_3)_3$(aq); $FeCl_3$(aq); FeS; $Fe(OH)_3$; $[Fe(NCS)]^{2+}$(aq); $Mn(NO_3)_2$(aq); MnO_2; MnO_4^-(aq); $Al(NO_3)_3$(aq); $Al(OH)_3$; $Cr(NO_3)_3$(aq); $CrCl_3$(aq); CrO_4^{2-}(aq); $PbCrO_4$; $Zn(NO_3)_2$(aq); ZnS; $K_2Zn[Fe(CN)_6]$.

69. Select reagents that will separate the members of the following pairs in one step, and write a skeletal equation for each separation. Fe^{3+} and Co^{2+}; Ni^{2+} and Al^{3+}; FeS and $Al(OH)_3$; FeS and CoS; MnS and NiS; Fe^{3+} and Cr^{3+}; $Cr(OH)_3$ and $Zn(OH)_2$; $Al(OH)_3$ and $Ni(OH)_2$; $Al(OH)_3$ and $Zn(OH)_2$; $Fe(OH)_3$ and MnO_2; MnO_2 and $Cr(OH)_3$.

70. (a) If a significant amount of iron is found in the test for cobalt, what does this suggest about your technique in Step B? (b) Refer to Table 33-1 and compare solubility products for FeS and ZnS. What is the likely fate of zinc ions?

71. Explain why the addition of an excess of potassium fluoride allows you to detect cobalt even if a large amount of iron is present when you test for cobalt.

*72. Oximes are very weakly acidic. We wrote the formula for dimethylglyoxime in simplified form as H_2DMG to emphasize this fact. The test for nickel is always done in a solution that contains a slight excess of aqueous NH_3. Suggest a reason why the test for nickel fails in solutions that contain a large excess of aqueous NH_3. [Hint: The formula for the red precipitate is $Ni(HDMG)_2$.]

73. What is the fate of zinc ions if insufficient aqueous NH_3 is added in Step J?

74. What problem is likely to result from prolonged boiling of the $NaOH/H_2O_2$ solution (Step F) when Al^{3+} ions are absent?

75. What concentration of sulfide ions is necessary to initiate the precipitation of (a) ZnS in a solution that is 0.050 M in $Zn(NO_3)_2$? (b) MnS in a solution that is 0.050 M in $Mn(NO_3)_2$?

76. What is the minimum pH necessary to initiate the precipitation of (a) $Mn(OH)_2$ in a solution that is 0.050 M in $Mn(NO_3)_2$? (b) $Al(OH)_3$ in a solution that is 0.050 M in $Al(NO_3)_3$?

Analysis of
Cation Group IV

Sea shells are mostly $CaCO_3$, with traces of transition metal ions.

Charles D. Winters

34-1 INTRODUCTION

The ions of analytical Group IV are referred to as the **insoluble carbonate group.** They are precipitated as carbonates from a buffered aqueous ammonia solution by the addition of ammonium carbonate.

Group IV contains the ions of calcium, strontium, and barium—three metals in Group IIA in the periodic table. They exhibit the +2 oxidation state in all their common compounds. Because they are derived from metals in the same family of the periodic table, these cations have similar properties. Therefore, separations of the individual ions are more difficult than separations in the earlier groups of cations. Magnesium ions are not precipitated in analytical Group IV because $MgCO_3$ is fairly soluble, and the aqueous ammonia solution is buffered to prevent precipitation of $Mg(OH)_2$, $K_{sp} = 1.5 \times 10^{-11}$ (Section 20-5 in the textbook).

Throughout this supplement, references to chapters or sections prior to Chapter 29 refer to *General Chemistry* by Whitten, Davis, Peck, and Stanley.

34-2 SOLUBILITIES OF COMPOUNDS OF CALCIUM, STRONTIUM, AND BARIUM

The sulfides and hydroxides of Ca^{2+}, Sr^{2+}, and Ba^{2+} ions are all fairly soluble compounds, so these cations are not precipitated in analytical Group I, II, or III. Because Ca^{2+}, Sr^{2+}, and Ba^{2+} ions have noble gas electron configurations and no partially filled d sublevels, most of their compounds are white.

Table 34-1 shows that the analytical Group IV carbonates are not nearly as insoluble as most of the compounds formed in earlier group precipitations and that the solubilities of

Most aqueous solutions that contain no cations other than Ca^{2+}, Sr^{2+}, and Ba^{2+} ions are colorless.

TABLE 34-1	Solubility Products and Molar Solubilities of Group IV Carbonates	
Compound	K_{sp}	**Molar Solubility**
$CaCO_3$	4.8×10^{-9}	$6.9 \times 10^{-5}\ M$
$SrCO_3$	9.4×10^{-10}	$3.1 \times 10^{-5}\ M$
$BaCO_3$	8.1×10^{-9}	$9.0 \times 10^{-5}\ M$

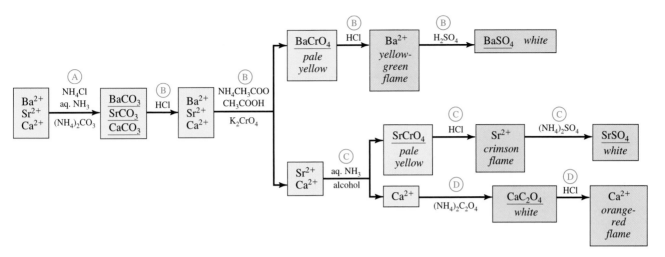

Figure 34-1 Analytical Group IV flow chart. Circled letters refer to analytical steps. Species in confirmatory tests are shown with blue backgrounds. Precipitated species are underlined.

TABLE 34-2	Solubility Products for Analytical Group IV Chromates, Sulfates, and Oxalates		
M^{2+}	$MCrO_4$	MSO_4	MC_2O_4
Ca^{2+}	7.1×10^{-4}	2.4×10^{-5}	2.3×10^{-9}
Sr^{2+}	3.6×10^{-5}	2.8×10^{-7}	5.6×10^{-8}
Ba^{2+}	2.0×10^{-10}	1.08×10^{-10}	1.1×10^{-7}

the Group IV carbonates are quite similar. As is often the case with solubilities, the solubilities of these carbonates do not correlate with the positions of the metals in the periodic table.

As the analytical Group IV flow chart in Figure 34-1 shows, Ba^{2+} ions are isolated by precipitating barium chromate, $BaCrO_4$, followed by reprecipitation as $BaSO_4$; Sr^{2+} ions are isolated as strontium chromate, $SrCrO_4$, followed by reprecipitation as strontium sulfate, $SrSO_4$; Ca^{2+} ions are isolated as calcium oxalate, CaC_2O_4.

Table 34-2 shows that the solubilities of the chromates, sulfates, and oxalates of the Group IV cations vary little. Therefore, it is relatively easy to precipitate compounds that *should* contain only one kind of cation but that are, in fact, contaminated with another kind of cation. Because these compounds contain one cation and one anion per formula unit, their solubility products can be compared. In each case the molar solubility is $\sqrt{K_{sp}}$.

Many oxalates precipitate as hydrated compounds, $CaC_2O_4 \cdot H_2O$, $SrC_2O_4 \cdot H_2O$, and $BaC_2O_4 \cdot 2H_2O$.

34-3 FLAME TESTS

Laboratory burner flames are hot enough to promote electrons in atoms and ions to higher energy levels. As these excited atoms and ions move out of the hot region of the flame, electrons drop back to their ground state energy levels (Figure 34-2). In some atoms and ions, energy is emitted as visible light (Section 5-12 in the textbook). Such atoms and ions impart

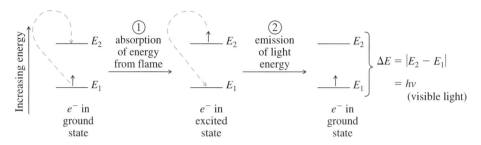

Figure 34-2 Electronic transitions in a flame test.

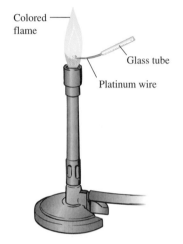

Figure 34-3 An illustration of how a flame test is performed.

characteristic colors to flames. The differences between energy levels are *very specific* and correspond to light of *specific wavelengths in the visible region* (and therefore to specific, characteristic colors). Calcium, strontium, and barium ions impart colors that are used as confirmatory tests for the Group IV cations, as do sodium and potassium ions in Group V. Metal chlorides are more volatile than other common salts, so flame tests are usually done on chloride solutions.

Flame tests are performed with a piece of platinum or nichrome wire sealed into the end of a piece of glass tubing or stuck into a small cork. Flame tests are very sensitive. The wire must be clean before a flame test is done. The end of the wire is bent into a small loop so that, when it is dipped into a solution, a film of liquid covers the loop. The burner flame is adjusted so that it is as hot as possible and there is a well-defined blue cone with very little color in the outer part of the flame. The wire loop is dipped in 6 M HCl (in a small test tube), brought slowly up to the outer edge of the blue cone, and held there until the loop is "red hot" (Figure 34-3). The hot loop is dipped in 6 M HCl, and the operation is repeated until the wire imparts no color to the flame.

The clean wire is dipped into the solution to be flame-tested and then brought up to the outer edge of the blue cone. The cations of Groups IV and V color flames as shown in Figure 34-4.

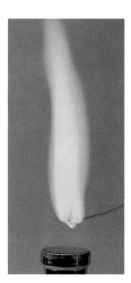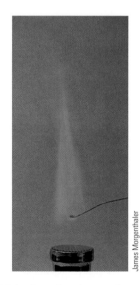

Figure 34-4 Flame colors for Groups IV and V cations. *Left to right:* barium (*apple green*), strontium (*crimson*), calcium (*brick red*), sodium (*yellow*), and potassium (*violet*).

The color imparted to the flame must be observed carefully. Different people see colors somewhat differently. Therefore, it is extremely important that *you* know how each color appears to *you*. Practice with known solutions until you know how each color appears. Never rely on your memory—always flame-test a known solution just before you flame-test an unknown. (Always clean the wire first by dipping it into 6 *M* HCl and then heating the wire.) Precipitates should always be dissolved in a small amount of HCl to provide solutions for flame tests.

One drop of 6 *M* HCl should be added to a few drops of a known or unknown solution before a flame test is performed.

34-4 PRECIPITATION OF ANALYTICAL GROUP IV CATIONS

Known and unknown solutions of the Group IV cations are either nitrates that contain a small amount of HNO_3 or chlorides that contain a little HCl.

The Group IV precipitating reagent, ammonium carbonate, contains the cation of a weak base and the anion of a weak polyprotic acid. Both hydrolyze.

$$NH_4^+ + H_2O \rightleftharpoons NH_3 + H_3O^+ \qquad K_a = 5.6 \times 10^{-10}$$

$$CO_3^{2-} + H_2O \rightleftharpoons HCO_3^- + OH^- \qquad K_b = 2.1 \times 10^{-4}$$

The carbonate ion hydrolyzes to a much greater extent than the ammonium ion, and therefore solutions of $(NH_4)_2CO_3$ do not contain sufficiently high concentrations of CO_3^{2-} ions to precipitate the Group IV carbonates completely. The Group IV precipitation is carried out in a buffered aqueous ammonia solution to suppress hydrolysis of the carbonate ion. The buffering action of NH_4Cl also suppresses the ionization of aqueous ammonia and prevents the precipitation of $Mg(OH)_2$ in Group IV.

Solutions from which Groups I, II, and III have been removed contain high concentrations of ammonium ions because aqueous ammonia was added in earlier procedures. Such solutions must be evaporated to dryness and heated quite strongly to decompose ammonium salts. Ammonium salts undergo thermal decomposition to produce the acid from which they were formed plus ammonia. Oxidizing acids, particularly HNO_3, may oxidize ammonia at elevated temperatures.

$$NH_4Cl(s) \xrightarrow{\text{heat}} NH_3(g) + HCl(g)$$

$$NH_4NO_3(s) \xrightarrow{\text{heat}} NH_3(g) + HNO_3(g)$$

$$NH_3(g) + HNO_3(g) \xrightarrow{\text{heat}} N_2O(g) + 2H_2O(g)$$

After all ammonium salts have been decomposed, the NH_4^+ concentration can be adjusted to the desired value. One drop of 12 *M* HCl is added and then neutralized by 6 *M* aqueous ammonia,

$$H^+ + Cl^- + NH_3 \longrightarrow NH_4^+ + Cl^-$$

after which the desired excess of aqueous ammonia is added. The precipitating reagent, 1 *M* $(NH_4)_2CO_3$, is then added to the carefully buffered solution.

The reactions by which the Group IV cations are precipitated can be represented as

$$Ca^{2+} + CO_3^{2-} \longrightarrow CaCO_3(s) \qquad \text{white}$$

$$Sr^{2+} + CO_3^{2-} \longrightarrow SrCO_3(s) \qquad \text{white}$$

$$Ba^{2+} + CO_3^{2-} \longrightarrow BaCO_3(s) \qquad \text{white}$$

Flame tests for strontium (*red*) and barium (*green*).

James Morgenthaler

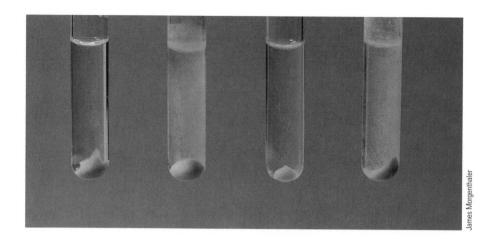

The Group IV precipitates. *Left to right:* BaCO₃, SrCO₃, CaCO₃, and a mixture of all three.

These carbonates precipitate as dense, fairly crystalline compounds, so there appears to be relatively little of the carbonate precipitate.

Step A

The first step in the analysis of Cation Group IV is the precipitation of the carbonates of Ba^{2+}, Sr^{2+}, and Ca^{2+}.

(1) If the solution to be analyzed was obtained from the Group III separation, add 3 drops of 16 M HNO_3 and evaporate to dryness in a casserole or crucible. Heat the casserole or crucible red hot for 5 min. Allow it to cool, and then dissolve the residue in 12 drops of H_2O and 1 drop of 12 M HCl. Centrifuge, and transfer the solution to a test tube.

(2) If the solution to be analyzed contains only Group IV cations, place 12 drops of the known or unknown in a test tube and add 1 drop of 12 M HCl.

To the solution resulting from either (1) or (2), add 6 M aqueous NH_3 dropwise until the solution is just basic. *Test after each drop of aqueous NH_3 is added.* Now add 1 drop of 6 M aqueous NH_3 in excess and 3 drops of 1 M $(NH_4)_2CO_3$. Mix well and place in the hot water bath at 70–80°C for 3 min. Cool the mixture and add 1 drop of 1 M $(NH_4)_2CO_3$ to test for complete precipitation. Separate the mixture when precipitation is complete. Treat the residue according to Step B. Save the liquid for the Group V analysis if you are analyzing a general unknown. Otherwise, discard it.

At temperatures above 80°C, $(NH_4)_2CO_3$ decomposes into ammonia, carbon dioxide, and water.

34-5 SEPARATION AND IDENTIFICATION OF BARIUM IONS

Metal carbonates are dissolved by acids that are stronger than carbonic acid, H_2CO_3 ($K_1 = 4.2 \times 10^{-7}$, $K_2 = 4.8 \times 10^{-11}$). Because the separation of Ba^{2+} ions from Sr^{2+} and Ca^{2+} ions is accomplished best in weakly acidic solutions, the carbonates are dissolved in HCl, the excess HCl is neutralized with aqueous NH_3, and the solution is then made slightly acidic with CH_3COOH. The four equilibria involved in the dissolution may be summarized as follows for calcium carbonate. The equilibria for $SrCO_3$ and $BaCO_3$ are similar.

$$CaCO_3(s) \rightleftharpoons Ca^{2+} + CO_3^{2-}$$
$$H^+ + CO_3^{2-} \rightleftharpoons HCO_3^-$$
$$HCO_3^- + H^+ \rightleftharpoons H_2CO_3$$
$$H_2CO_3 \rightleftharpoons CO_2(g) + H_2O$$

Heating the reaction mixture decreases the solubility of CO_2 in water and shifts these equilibria to the right so that the carbonates dissolve completely. The overall reaction may be represented as

$$2H^+ + CaCO_3(s) \longrightarrow Ca^{2+} + CO_2(g) + H_2O$$

Table 34-2 shows that barium chromate is less soluble than strontium chromate and that strontium chromate is less soluble than calcium chromate. These differences in solubilities provide the basis for separating barium ions from strontium and calcium ions.

$$Ba^{2+} + CrO_4^{2-} \longrightarrow BaCrO_4(s) \qquad \text{pale yellow}$$

However, strontium chromate tends to coprecipitate with barium chromate, so the chromate ions are added slowly to a hot solution to minimize this probability. Strontium chromate forms supersaturated solutions readily, and keeping the solution hot increases the probability that $SrCrO_4$ will remain in solution.

The concentration of Sr^{2+} ions is approximately $0.040\ M$ in the solution from which $BaCrO_4$ is precipitated. The maximum concentration of CrO_4^{2-} ions that can be present without exceeding K_{sp} for $SrCrO_4$ can be calculated.

Coprecipitation is the phenomenon in which a small amount of a compound precipitates *with* a similar compound whose K_{sp} is exceeded.

$$[Sr^{2+}][CrO_4^{2-}] = 3.6 \times 10^{-5}$$

$$[CrO_4^{2-}] = \frac{3.6 \times 10^{-5}}{[Sr^{2+}]} = \frac{3.6 \times 10^{-5}}{4.0 \times 10^{-2}} = 9.0 \times 10^{-4}\ M$$

Therefore, it is necessary to keep the concentration of CrO_4^{2-} ions below $9.0 \times 10^{-4}\ M$. The concentration of Ba^{2+} ions that remains in such a solution, that is, unprecipitated, can also be calculated.

$$[Ba^{2+}][CrO_4^{2-}] = 2.0 \times 10^{-10}$$

$$[Ba^{2+}] = \frac{2.0 \times 10^{-10}}{[CrO_4^{2-}]} = \frac{2.0 \times 10^{-10}}{9.0 \times 10^{-4}} = 2.2 \times 10^{-7}\ M$$

This calculation tells us that Ba^{2+} ions are almost completely precipitated from solutions that are $9.0 \times 10^{-4}\ M$ in CrO_4^{2-} ions.

Chromate ions are converted to dichromate ions in acidic solutions.

$$2CrO_4^{2-} + 2H^+ \rightleftharpoons Cr_2O_7^{2-} + H_2O$$

Therefore, we can control the concentrations of CrO_4^{2-} and $Cr_2O_7^{2-}$ in solutions by varying the concentration of H^+. The equilibrium constant for the above reaction is

$$\frac{[Cr_2O_7^{2-}]}{[CrO_4^{2-}]^2[H^+]^2} = 4.2 \times 10^{14}$$

The concentration of $Cr_2O_7^{2-}$ is about $0.18\ M$ in the solution from which $BaCrO_4$ is precipitated. We can solve the equilibrium constant expression for the $[H^+]$ that will maintain $[CrO_4^{2-}]$ at $9.0 \times 10^{-4}\ M$.

The confirmatory tests for barium: the flame test and precipitated BaSO₄.

$$[H^+]^2 = \frac{[Cr_2O_7^{2-}]}{[CrO_4^{2-}]^2(4.2 \times 10^{14})} = \frac{0.18}{(9.0 \times 10^{-4})^2(4.2 \times 10^{14})} = 5.3 \times 10^{-10}$$

$$[H^+] = 2.3 \times 10^{-5}\,M \longleftarrow \text{minimum value for } [H^+]$$

Therefore, to precipitate $BaCrO_4$ almost completely while leaving Sr^{2+} ions in solution, the concentration of H^+ ions should be $2.3 \times 10^{-5}\,M$ or slightly higher (about pH 4.5) to provide some margin of safety.

The ionization constant for CH_3COOH is 1.8×10^{-5}. Buffer solutions (Section 18-7 in the textbook) containing CH_3COOH and NH_4CH_3COO are used to control the pH of the solution in which $BaCrO_4$ is precipitated.

$$\frac{[H^+][CH_3COO^-]}{[CH_3COOH]} = 1.8 \times 10^{-5}$$

$$\frac{[CH_3COO^-]}{[CH_3COOH]} = \frac{1.8 \times 10^{-5}}{[H^+]} = \frac{1.8 \times 10^{-5}}{2.3 \times 10^{-5}} = 0.78$$

The ratio of [salt] to [acid] should be 0.78 or slightly less to provide some margin of safety.

After $BaCrO_4$ has been precipitated and isolated, it must be washed to remove traces of Sr^{2+}, Ca^{2+}, and K^+ ions (from K_2CrO_4) that interfere with the flame test. It is then dissolved in dilute HCl, and a flame test is performed on the resulting solution.

$$2BaCrO_4(s) + 2H^+ \rightleftharpoons 2Ba^{2+} + Cr_2O_7^{2-} + H_2O$$

The high concentration of H^+ reduces the concentration of CrO_4^{2-} to the point that $[Ba^{2+}][CrO_4^{2-}] < K_{sp}$ and $BaCrO_4$ dissolves.

After the flame test has been performed, the solution is treated with H_2SO_4. The formation of a highly crystalline, nearly colorless precipitate, $BaSO_4$, is taken as the final confirmatory test for barium ions.

$$Ba^{2+} + SO_4^{2-} \longrightarrow BaSO_4(s) \qquad \text{white}$$

Step B

Use the precipitate from Step A, containing $BaCO_3$, $CaCO_3$, and $SrCO_3$. Add 10 drops of water and 2 drops of 6 M HCl to the precipitate, stir well, and place the test tube in the hot water bath for 3 min. All of the white precipitate should dissolve. Stir several times to ensure the removal of CO_2. Add 2 drops of 6 M aqueous ammonia, mix well, and then add 2 drops of 6 M CH_3COOH. Mix well and place the test tube in the hot water bath for 2 min. Use a clean *capillary pipet that delivers small drops* to add 1 *tiny* drop of 0.1 M K_2CrO_4 to the hot solution. Mix well and place in the hot water bath for 1 min. Repeat this step until 10 *tiny* drops of 0.1 M K_2CrO_4 have been added, and then heat the mixture for 3 min with frequent stirring. A finely divided pale yellow precipitate indicates the presence of barium. Separate the mixture and save the solution for Step C.

Wash the precipitate twice with 3 drops of 3 M aqueous ammonia and discard the wash liquid. Dissolve the precipitate in 1 drop of 6 M HCl and 2 drops of water, and then perform a flame test on the resulting solution. Barium ions give a short-lived yellow-green color to the flame. Add 1 drop of 6 M H_2SO_4 to the solution used for the flame test. A finely divided, nearly colorless precipitate, $BaSO_4$, confirms the presence of barium ions.

34-6 SEPARATION AND IDENTIFICATION OF STRONTIUM IONS

The solubility product for $SrCrO_4$, 3.6×10^{-5}, is fairly large, and so a high concentration of CrO_4^{2-} is required to exceed K_{sp} and precipitate $SrCrO_4$. The addition of aqueous ammonia to the CH_3COOH/NH_4CH_3COO buffer solution neutralizes CH_3COOH and shifts the following equilibrium to the left.

$$2CrO_4^{2-} + 2H^+ \rightleftharpoons Cr_2O_7^{2-} + H_2O$$

Recall that CrO_4^{2-} ions are yellow and $Cr_2O_7^{2-}$ ions are orange.

When the concentration of CrO_4^{2-} is sufficiently high that K_{sp} for $SrCrO_4$ is exceeded, precipitation should occur. However, $SrCrO_4$ forms supersaturated solutions readily. Precipitation can usually be induced by the addition of ethyl alcohol, which decreases the polarity of the solvent and the solubility of $SrCrO_4$.

$$Sr^{2+} + CrO_4^{2-} \longrightarrow SrCrO_4(s) \qquad \text{pale yellow}$$

After $SrCrO_4$ has been isolated, it must be washed to remove traces of Ca^{2+} and K^+ ions that would interfere with the flame test.

Step C

Use the solution from Step B, containing Sr^{2+} and Ca^{2+}. Add 6 M aqueous ammonia dropwise, stirring after each drop, until the orange solution becomes a pretty pastel yellow. No more than 4 drops should be required. Heat the solution in the water bath for 3 min, cool it to room temperature, and add enough alcohol to double the volume of the solution. Mix well, and allow the mixture to stand for 3 min with frequent stirring. A finely divided yellow precipitate indicates the presence of strontium. Separate the mixture and save the solution for Step D.

Wash the precipitate twice, using a mixture of 3 drops of alcohol and 2 drops of 3 M aqueous ammonia each time. *Add the alcohol first.* Discard the wash solutions. Add 1 drop of 6 M HCl and 2 drops of water to dissolve the precipitate. Flame-test for Sr^{2+}. Strontium ions impart a crimson (pink-red) color to the flame. The flame test is the confirmatory test for Sr^{2+} ions.

If you are unable to decide whether Sr^{2+} ions are present, evaporate the solution that was flame-tested until all the liquid has disappeared. Do not bake. Cool, add 10 drops of water, mix well, and centrifuge. Transfer the liquid to a clean test tube and add 5 drops of 2.0 M $(NH_4)_2SO_4$. Mix well and allow the solution to stand for 5 min. A nearly colorless crystalline precipitate, $SrSO_4$, demonstrates the presence of Sr^{2+} *unless* Ba^{2+} ions were incompletely removed earlier.

James Morgenthaler

The confirmatory test for strontium.

34-7 SEPARATION AND IDENTIFICATION OF CALCIUM IONS

Calcium oxalate, CaC_2O_4, $K_{sp} = 2.3 \times 10^{-9}$, is precipitated from solutions that contain calcium ions by the addition of ammonium oxalate, $(NH_4)_2C_2O_4$.

$$Ca^{2+} + C_2O_4^{2-} \longrightarrow CaC_2O_4(s) \qquad \text{white}$$

The solution in which CaC_2O_4 is precipitated contains potassium ions (from the addition of K_2CrO_4). The precipitate must be washed to remove potassium ions before a flame test can be performed. The precipitate is dissolved in hydrochloric acid, and the flame test is performed on the resulting solution.

The confirmatory test for calcium.

$$CaC_2O_4(s) + H^+ \rightleftharpoons Ca^{2+} + HC_2O_4^-$$

Calcium ions impart an orange-red (brick red) color to the flame.

Step D

Use the solution from Step C, containing Ca^{2+}. Place the solution in the hot water bath for 3 min, add 3 drops of 0.50 M $(NH_4)_2C_2O_4$, and mix well. If calcium ions are present, a white crystalline precipitate, CaC_2O_4, should form. Because CaC_2O_4 forms supersaturated solutions, it may be necessary to scratch the walls of the test tube with a stirring rod to initiate precipitation. Cool the mixture, separate it, and discard the liquid.

Wash the precipitate twice, using 6 drops of water and 2 drops of 6 M aqueous ammonia each time. Discard the wash liquid. Add 1 drop of 6 M HCl and 2 drops of water to the precipitate and perform a flame test. An orange-red flame confirms the presence of calcium ions.

Now that you have completed the Group IV unknown, flame-test the original solution and compare the results with those obtained using known solutions. Unknowns that contain only a single ion give its characteristic color to the flame. The presence of Ba^{2+} and Sr^{2+}, or Ba^{2+} and Ca^{2+}, in an unknown can be detected easily. Detecting the presence of Sr^{2+} and Ca^{2+} in the same unknown is a little more difficult. The unambiguous detection of all three Group IV cations in a single unknown is even more difficult. However, the results of the tests you have performed for the individual ions must be compatible with a flame test on the original Group IV unknown.

Exercises

General Questions on Analytical Group IV

1. In each of the previous groups, we have studied metals that exhibit variable oxidation states in their compounds. The metals in Group IV do not. Why?
2. Why are most compounds of the Group IV cations white in the solid state and colorless in aqueous solutions?
3. What is the solubility rule for metal carbonates? What problems would be caused by the incomplete precipitation of Group III cations in a general unknown?
4. (a) In general unknowns, the centrifugate from the Group III precipitation is always evaporated to dryness with nitric acid. Why? (b) Why is this step unnecessary for unknowns that contain only Group IV cations?
5. (a) For which Group IV carbonate is the molar solubility highest? lowest? (b) What is the ratio (highest molar solubility)/(lowest molar solubility) for the carbonates in (a)? What does this tell you?
6. (a) What is the basis for flame tests? (b) Why are flame tests very specific? (c) Why should one always observe the color of a "known" flame test just before the unknown is flame-tested? (d) List the colors imparted to laboratory burner flames by Group IV cations.

Analytical Group IV Reactions

Write balanced net ionic equations for reactions that occur when the following are mixed. Indicate colors of all precipitates.

7. Barium nitrate and ammonium carbonate in buffered aqueous ammonia.
8. Strontium nitrate and ammonium carbonate in buffered aqueous ammonia.
9. Calcium nitrate and ammonium carbonate in buffered aqueous ammonia.
10. Barium carbonate dissolves in 1 M HCl.
11. Strontium carbonate dissolves in 1 M HCl.
12. Calcium carbonate dissolves in 1 M HCl.
13. Barium acetate and potassium chromate (slightly acidic solution).
14. Barium chromate dissolves in 2 M HCl.
15. Dichromic acid reacts with *hot* 2 M HCl.
16. Barium chloride and dilute sulfuric acid.
17. Strontium acetate and potassium chromate.
18. Strontium chromate dissolves in 2 M HCl.
19. Strontium chloride and ammonium sulfate.
20. Calcium acetate and ammonium oxalate.
21. Calcium oxalate dissolves in 2 M HCl.

Other Questions and Problems

22. Why should the solution in which the Group IV precipitation is done be kept below 100°C?

23. Describe and write equations for the important equilibria in buffered aqueous ammonia solutions that contain ammonium carbonate.

24. (a) Describe and write equations for the important equilibria in the dissolution of Group IV carbonates in 1 M HCl. Use $SrCO_3$ as your example. (b) Why do you wait until the carbonates have dissolved before adding aqueous NH_3 in Step B? (c) What is the purpose of adding aqueous NH_3? (d) What are the likely consequences of adding too much aqueous NH_3 in Step B? insufficient aqueous NH_3?

25. (a) Calculate the molar solubilities of the Group IV chromates in pure water. (b) Let M_{E1} refer to molar solubilities of these chromates; calculate the following ratios: $(M_{Ba})/(M_{Sr})$, $(M_{Ba})/(M_{Ca})$, $(M_{Sr})/(M_{Ca})$. What do these ratios tell us?

26. The equilibrium constant for the reaction

$$2CrO_4^{2-} + 2H^+ \rightleftharpoons Cr_2O_7^{2-} + H_2O$$

is 4.2×10^{14}.

(a) Write the equilibrium constant expression. (b) What is the value of the equilibrium constant for the reaction

$$Cr_2O_7^{2-} + H_2O \rightleftharpoons 2CrO_4^{2-} + 2H^+?$$

(c) Which ion, CrO_4^{2-} or $Cr_2O_7^{2-}$, is present in greater concentration in a solution (pH = 7.00) in which 0.10 mol/L of $K_2Cr_2O_7$ was dissolved? (d) Repeat (c) for pH = 2.00 and pH = 12.00.

27. Wires made of nichrome, or similar alloys, are used for flame tests in many laboratories. Why do the HCl solutions used to clean nichrome wires turn green? Which metals are suggested by the name "nichrome"?

28. Explain the equilibria involved in the dissolution of $BaCrO_4$ in HCl as you prepare for the flame test.

29. Dichromate ions are strong oxidizing agents in acidic solutions, and they oxidize chloride ions to elemental chlorine in hot solutions. Even if you use a platinum wire for flame tests, the solution that is flame-tested for barium ions turns green rapidly after the hot wire is plunged into it a few times. Why?

30. Suppose you forget to wash the $BaCrO_4$ before you perform a flame test. What unfortunate consequence would you expect? (*Hint:* What reagent was added to precipitate $BaCrO_4$?)

31. Why is aqueous ammonia added to the solution from which $BaCrO_4$ has been removed before $SrCrO_4$ is precipitated?

32. (a) Why is alcohol added to the solution from which $SrCrO_4$ is to be precipitated? (b) What would be the result of adding too little alcohol? Too much alcohol?

33. Refer to the last sentence in Step C. Why do you suppose *unless* is italicized? (Refer to Table 34-2.)

35 Analysis of Cation Group V

Fireworks display colors emitted by excited metal ions.

35-1 INTRODUCTION

Cation Group V is known as the *soluble group* because its cations are not precipitated by the reagents used to precipitate and separate the first four groups or by any other single reagent. These cations are magnesium, Mg^{2+}, sodium, Na^+, potassium, K^+, and ammonium, NH_4^+. None of the metals in this group exhibits variable oxidation states.

The solubility guidelines tell us that all common *simple* compounds of Na^+, K^+, and NH_4^+ are soluble in water. (In analytical Group II we precipitated $MgNH_4AsO_4$, an insoluble double salt, in the confirmatory test for arsenic.) Some other complex compounds of these three cations are insoluble. Several compounds of magnesium are insoluble, so magnesium ions are easily separated from the other members of this group.

You should review the solubility guidelines in Section 4-2 and Table 4-8 of the textbook *General Chemistry* by Whitten, Davis, Peck, and Stanley.

> **AN IMPORTANT NOTE** The test for the ammonium ion, Step F, must always be done on the *original* unknown because aqueous ammonia is added at several points in the analytical procedures. Save some of the original unknown for Step F.
>
> Most of the unknown solutions are acidic, and gaseous ammonia may be absorbed from the air in the laboratory. Therefore, Group V and general unknowns should be kept tightly stoppered.

The solution to be analyzed may contain only analytical Group V ions, or it may be the centrifugate from the Group IV separation. In general unknowns, care must be exercised before the analysis of Group V is begun to remove any traces of Group IV cations that may have been carried into Group V. Such procedures are not necessary for solutions that contain *only* Group V cations.

Removal of Traces of Barium, Strontium, and Calcium Ions

Even small amounts of the Group IV cations give misleading results in the initial precipitation of magnesium ions. (Traces of Group IV cations fail to precipitate at appropriate points in the Group IV procedures when the acidity is not adjusted properly. This occurs when the concentration of ammonium ions is too high.) As a precautionary measure, a few drops each of $(NH_4)_2C_2O_4$ and $(NH_4)_2SO_4$ solutions are added to the solution to be analyzed. Any precipitate is separated and discarded before the analysis of Group V is begun. The $C_2O_4^{2-}$ ions precipitate traces of Ca^{2+} and Sr^{2+} ions, and the SO_4^{2-} ions precipitate traces of Ba^{2+} ions.

Halite crystals (naturally occurring NaCl).

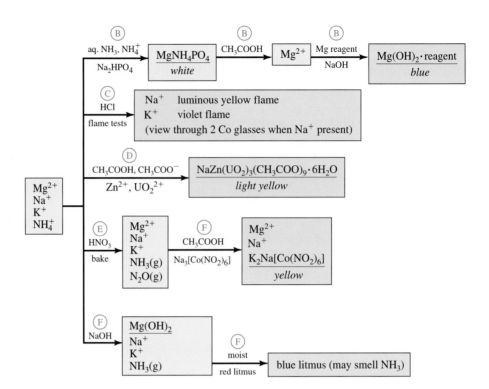

Figure 35-1 Group V flow chart. Circled letters refer to analytical steps. Species in confirmatory tests are shown with blue backgrounds. Precipitated species are underlined.

Step A

Use this procedure for general unknowns only; begin with Step B for knowns and unknowns that contain only Group V cations. Use the solution from the Group IV separation, containing Mg^{2+}, Na^+, K^+, NH_4^+, and possibly traces of Ca^{2+}, Sr^{2+}, and Ba^{2+}. Add 1 drop of 0.5 M $(NH_4)_2C_2O_4$ and 1 drop of 2 M $(NH_4)_2SO_4$ to the solution from which Group IV cations were removed, mix well, and centrifuge the mixture. Wrap a *tiny* piece of cotton around the end of a capillary pipet and draw the liquid into the pipet. Discard the piece of cotton and any precipitate. Save the solution for Step B.

Although Group V cations cannot be separated individually in a schematic way as was done in earlier groups, each exhibits properties that enable us to test for that ion individually, as Figure 35-1 shows. The reactions indicated in the flow chart are specific for each ion *under carefully controlled conditions*. In the hands of the inexperienced, the precipitation reactions for sodium and potassium ions are less reliable than the flame tests for these ions.

35-2 SEPARATION AND IDENTIFICATION OF MAGNESIUM IONS

A portion of the solution is tested for magnesium ions. The test for arsenate ions, AsO_4^{3-}, in analytical Group II involved precipitating magnesium ammonium arsenate, $MgNH_4AsO_4$. The formation of a similar compound, magnesium ammonium phosphate,

$MgNH_4PO_4$, is used to separate magnesium ions from the other Group V cations. The reaction of disodium hydrogen phosphate with magnesium ions in buffered aqueous ammonia solutions is

$$Mg^{2+} + NH_3 + HPO_4{}^{2-} \longrightarrow MgNH_4PO_4(s) \qquad \text{white}$$

The solid $MgNH_4PO_4$ is separated and dissolved in acetic acid, which reverses the above reaction by converting $PO_4{}^{3-}$ ions to $HPO_4{}^{2-}$ ions. This shifts the equilibrium to the left. Recall that K_3 for H_3PO_4 is only 3.6×10^{-13} (Appendix F). Therefore, even the weak acid CH_3COOH ($K_a = 1.8 \times 10^{-5}$) is strong enough to dissolve most phosphates. The concentrations of the ions are reduced so that $[Mg^{2+}][NH_4{}^+][PO_4{}^{3-}] < K_{sp}$, and the compound dissolves.

$$MgNH_4PO_4(s) + 2CH_3COOH \rightleftharpoons Mg^{2+} + NH_4{}^+ + H_2PO_4{}^- + 2CH_3COO^-$$

The addition of sodium hydroxide solution precipitates magnesium hydroxide, a white compound.

$$Mg^{2+} + 2OH^- \longrightarrow Mg(OH)_2(s)$$

When $Mg(OH)_2$ is precipitated in the presence of an organic dye known as magnesium reagent (*p*-nitrobenzeneazo)resorcinol, the purple dye is adsorbed by the white $Mg(OH)_2$, and the precipitate appears blue. It is sometimes referred to as a "blue lake."

Lake is a general term for a coprecipitate of an organic dye with a metallic hydroxide or salt. This kind of reaction can be used to dye cloth.

Step B

Use 6 drops of the solution from Step A or 12 drops of the Group V known or unknown solution, containing Mg^{2+}, Na^+, K^+, and $NH_4{}^+$. Test the solution with litmus. If it is not basic, add 6 M aqueous ammonia dropwise, with stirring, until it becomes basic and then add 1 drop in excess. Add 2 drops of 1.5 M Na_2HPO_4 and mix well, scratching the walls of the test tube with a stirring rod to initiate precipitation if necessary. The appearance of a white crystalline precipitate, $MgNH_4PO_4$, proves the presence of magnesium ions. Centrifuge and separate the mixture. Add 1 drop of 6 M CH_3COOH and 4 drops of water to the precipitate. Mix well; after all the precipitate has dissolved, add 1 drop of magnesium reagent (not to be confused with magnesia mixture). Mix well, add 5 drops of 4 M NaOH, stir, and centrifuge the mixture. The presence of a *sky blue precipitate* proves that magnesium ions are present. Observe carefully because the magnesium reagent makes the solution purple.

The confirmatory test for magnesium, $MgNH_4AsO_4$.

35-3 FLAME TESTS FOR SODIUM AND POTASSIUM IONS

The most reliable tests for sodium and potassium ions are the flame tests. Both impart characteristic colors to the burner flame. Traces of sodium contaminate almost everything. Use caution in reporting sodium ions.

Flame tests may be performed on the solution obtained in Step A (removal of traces of Group IV cations) or on the original Group V known and unknown solutions. Sodium ions impart an intense yellow color to the burner flame. Flame tests on a known solution should always be compared with those done on unknowns.

Potassium ions impart a characteristic violet color to the flame. However, when sodium ions are also present, their intense yellow color completely masks the violet color produced

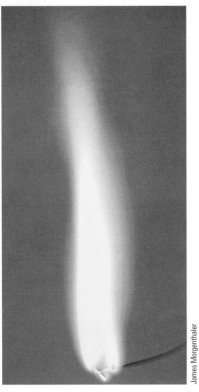

A confirmatory test for sodium.

by potassium ions. Two thicknesses of cobalt glass filter out the yellow color produced by sodium ions so that the potassium flame test can be observed even when sodium ions are present. You should practice observing the potassium flame test through two pieces of cobalt glass until you are comfortable with it. First practice on a solution that contains potassium ions, and then use a solution that contains both sodium and potassium ions. Neither magnesium ions nor ammonium ions impart color to laboratory burner flames.

CAUTION!

Detergents contain large amounts of sodium compounds. Therefore, all test tubes and pipets that are used in the tests for sodium ions must be rinsed thoroughly to remove traces of sodium compounds.

Step C
Use 6 drops of the solution from Step A or 6 drops of the Group V known or unknown, containing Mg^{2+}, Na^+, K^+, and NH_4^+. Add 1 drop of 6 M HCl and flame-test the solution. Observe the flame to determine whether sodium ions are present. If sodium ions are present (even in trace amounts, as they are in almost everything), observe the flame through *two* thicknesses of cobalt glass to determine whether potassium ions are present. Do not confuse the red-hot wire with a flame test for potassium ions. An intense yellow flame and a characteristic violet flame prove the presence of sodium and potassium ions, respectively.

35-4 PRECIPITATION TEST FOR SODIUM IONS

Sodium ions can also be detected by precipitating a crystalline triple salt, sodium zinc uranyl acetate hexahydrate, $NaZn(UO_2)_3(CH_3COO)_9 \cdot 6H_2O$. A saturated solution of zinc acetate, $Zn(CH_3COO)_2$, and uranyl acetate, $UO_2(CH_3COO)_2$, in acetic acid is known as sodium reagent. Because Mg^{2+}, K^+, and NH_4^+ ions do not form precipitates with sodium reagent, the precipitation test may be performed in the presence of these cations.

$$Na^+ + Zn^{2+} + 3UO_2^{2+} + 9CH_3COO^- + 6H_2O \longrightarrow$$
$$NaZn(UO_2)_3(CH_3COO)_9 \cdot 6H_2O(s)$$
light yellow

The formation of this precipitate is often very slow. This procedure should be done as early in the laboratory period as possible.

Step D
If you use the solution from Step A, perform (a) and then (c). If you use the original Group V known or unknown, perform (b) and then (c). The solution contains Mg^{2+}, Na^+, K^+, and NH_4^+.
(a) To 3 drops of the solution from Step A, add 3 M CH_3COOH until it is acidic to litmus.
(b) To 3 drops of the Group V known or unknown solution, add 1 M aqueous ammonia until it is basic to litmus (test after each drop). Then add 1 drop of 3 M CH_3COOH.

(c) Place 10 drops of sodium reagent in a test tube and add 2 drops of the solution prepared in (a) or (b). Mix well and allow to stand for an hour if necessary. The appearance of a highly crystalline yellow precipitate indicates that sodium ions are present.

35-5 PRECIPITATION TEST FOR POTASSIUM IONS

After ammonium ions have been removed, potassium ions can be detected in the solution by formation of an insoluble yellow compound, dipotassium sodium hexanitrocobaltate(III), $K_2Na[Co(NO_2)_6]$. A saturated solution of sodium hexanitrocobaltate(III), $Na_3[Co(NO_2)_6]$, is added to a weakly acidic solution that contains potassium ions.

$$2K^+ + Na^+ + [Co(NO_2)_6]^{3-} \longrightarrow K_2Na[Co(NO_2)_6](s) \qquad \text{yellow}$$

However, ammonium ions form a similar precipitate, so they must be removed before the test for potassium ions can be done. Destruction of ammonium ions is accomplished by evaporation with nitric acid followed by ignition to dull red heat, which volatilizes any remaining ammonium salts. These reactions may be summarized in simplified form as

$$NH_4Cl(s) \xrightarrow{\text{heat}} NH_3(g) + HCl(g)$$

$$NH_4NO_3(s) \xrightarrow{\text{heat}} NH_3(g) + HNO_3(g)$$

$$NH_3(g) + HNO_3(g) \xrightarrow{\text{heat}} N_2O(g) + 2H_2O(g)$$

Heat all parts of the dish in which the evaporation and ignition are done so that the expulsion of ammonium compounds is complete.

If NH_4^+ ions are incompletely removed, all is not lost. The test tube that contains $K_2Na[Co(NO_2)_6]$ or $(NH_4)_2Na[Co(NO_2)_6]$ (or both) can be heated until the solution turns pink. The pink color shows that $[Co(NO_2)_6]^{3-}$ ions have decomposed and that Co(III) has been reduced to Co(II) by nitrite ions in the hot solution. Nitrite ions react with ammonium ions at 100°C and oxidize them to nitrogen and water.

$$NO_2^- + NH_4^+ \longrightarrow N_2 + 2H_2O$$

After any remaining NH_4^+ ions have been destroyed, more $Na_3[Co(NO_2)_6]$ can be added, and potassium ions can be detected.

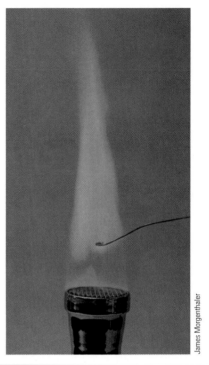

A confirmatory test for potassium.

James Morgenthaler

Step E

Use the remainder of the solution from Step A, containing Mg^{2+}, Na^+, K^+, and NH_4^+, or 10 drops of the original Group V known or unknown. Transfer the solution to a casserole (or a crucible or a small evaporating dish), add 5 drops of 16 M HNO_3 and then carefully evaporate the solution to dryness *under the hood*. Heat the casserole to a dull red heat for a few minutes. Direct the hot part of the flame at all parts of the *outer* surface of the casserole to ensure complete decomposition of ammonium compounds. Allow the casserole to cool to near room temperature and add 2 drops of 1 M HCl and 6 drops of 6 M CH_3COOH. Swirl the liquid so that it touches all the inner surface of the casserole and then *carefully* heat the solution to boiling. Add 4 drops of this solution to 10 drops of the saturated $Na_3[Co(NO_2)_6]$

solution and warm gently (50°C) for 5 min. A yellow precipitate indicates the presence of potassium ions if ammonium ions were removed completely.

If NH_4^+ ions were incompletely removed, $(NH_4)_2Na[Co(NO_2)_6]$ also precipitates at this point. Heat the solution and yellow precipitate in the hot water bath (100°C) until the solution turns pink. Cool the pink solution and add 3 more drops of $Na_3[Co(NO_2)_6]$. A yellow precipitate indicates conclusively that potassium ions are present.

35-6 TEST FOR AMMONIUM IONS

The test for ammonium ions must always be done on the original known or unknown solution, because aqueous ammonia is added in several procedures. Aqueous ammonia is a weak base, and it ionizes only slightly.

$$NH_3 + H_2O \rightleftharpoons NH_4^+ + OH^-$$

The addition of ammonium ions to a solution of a strong base such as NaOH results in the formation and liberation of gaseous ammonia, that is, the reverse of the preceding reaction.

$$NH_4^+ + OH^- \text{ (excess)} \longrightarrow NH_3(g) + H_2O$$

The liberated ammonia can be detected by its characteristic odor or by contact with an acid–base indicator such as litmus.

The confirmatory test for ammonium ions.

Step F

Detection of the ammonium ion. Place a piece of red litmus paper on the convex side of a clean watch glass and wet the paper with distilled water so that it adheres to the watch glass. Place 20 drops of 4 *M* NaOH in a small beaker (or a small evaporating dish) supported on a ring stand. Heat the beaker *gently* until it is warm to the touch, but do *not* boil the NaOH solution. Add 5 drops of the *original* known or unknown solution to the warm NaOH. Place the watch glass over the beaker or evaporating dish immediately and watch for a color change (red litmus turns blue), which proves the presence of ammonium ions. You may wish to remove the watch glass and smell the ammonia.

A word of caution. The NaOH solution must not be hot enough to boil—litmus always turns blue when NaOH spatters onto it!

Step G

A final check. A flame test should be performed on a few drops of the original solution to which 1 drop of 6 *M* HCl has been added. The results should be consistent with those obtained for the individual ions.

Keep in mind the fact that, in general, the flame tests for unknowns can be observed for any Group IV ions that are present *only if* sodium ions are absent, because sodium ions obscure the flame tests for Group IV cations. Copper(II) ions impart an intense green color to the flame, so the results of a flame test may not be very meaningful in unknowns that contain copper.

Exercises

General Questions on Analytical Group V

1. (a) List the cations in Group V. (b) Why is Group V called the soluble group?
2. Why are ammonium sulfate and ammonium oxalate added to the solution from which the Group IV cations were removed (Step A) before testing for Group V cations?
3. Why is the test for the ammonium ion always performed on a sample of the original unknown solution?
4. (a) Why does litmus paper often change colors when exposed to the air in the laboratory for long periods of time? (b) What compounds are likely to be responsible for the color changes?

Analytical Group V Reactions

Write balanced net ionic equations for reactions that occur when the following are mixed. Indicate colors of all precipitates.

5. Ammonium nitrate + 4 M NaOH.
6. Magnesium nitrate + disodium hydrogen phosphate in buffered aqueous NH_3.
7. Magnesium ammonium phosphate + dilute CH_3COOH.
8. Magnesium acetate + 4 M NaOH.
9. The thermal decomposition of ammonium chloride.
10. The thermal decomposition of ammonium nitrate (two reactions).
11. Potassium acetate + sodium hexanitrocobaltate(III) in excess CH_3COOH.
12. Potassium nitrite + ammonium acetate at 100°C in excess CH_3COOH.
13. Sodium nitrate + zinc acetate + uranyl acetate in excess CH_3COOH.

Other Questions and Problems

14. In the precipitation test for potassium (Step E) you are told to heat the solution that contains the yellow precipitate until the solution turns pink. Why?
15. In the alternative confirmatory test for cobalt [Step E(2) in Chapter 33], we represented the insoluble complex compound as $K_3[Co(NO_2)_6]$. In the precipitation test for potassium, we represented the insoluble compound as $K_2Na[Co(NO_2)_6]$. Can you suggest a reason? (*Hint:* What reagent did we add to test for cobalt? In what concentration?)
16. (a) Why are flame tests so distinctive? (b) What is the function of cobalt glass in the flame test for potassium?
17. If you were given a sample of a white solid known to be one of the following, how would you determine which one it is? Answer the question so that you could identify each absolutely. $MgSO_4$, Na_2SO_4, K_2SO_4, $(NH_4)_2SO_4$.
18. Assume that you are given a solution that contains only one Group V cation. How could you determine which cation is present, using *only* two tests? Explain.

36 Ionic Equilibria in Qualitative Analysis

OUTLINE

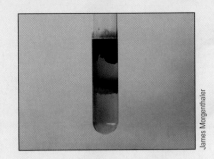

Some equilibria involving Co^{2+} ions.

In Chapters 18, 19, and 20 of the textbook we presented the fundamental ideas about equilibria in aqueous solutions. We now build upon those concepts to explain some of the more sophisticated equilibria encountered in qualitative analysis.

36-1 DISSOLVING PRECIPITATES

In our laboratory work we often find it necessary to dissolve precipitates. A precipitate dissolves when the concentrations of its ions are reduced so that K_{sp} is no longer exceeded, that is, when $Q_{sp} < K_{sp}$. Precipitates can be dissolved by the three types of reactions discussed in Section 20-6 in the textbook. All involve removing ions from solution. Please review Section 20-6 carefully.

The cations in many slightly soluble compounds can form complex ions. This often results in dissolution of the slightly soluble compound. From the discussion of complex ion formation in Section 20-6 in the textbook, we see that the more effectively a ligand competes for a coordination site on the metal ions, the smaller K_d is. This tells us that, in a comparison of complexes with the same number of ligands, the smaller the K_d value, the more stable the complex ion. Some complex ions and their dissociation constants, K_d, are listed in Appendix I.

Solubility products, like other equilibrium constants, are thermodynamic quantities. They tell us nothing about how fast a given reaction occurs, only that it can, or cannot, occur under specified conditions.

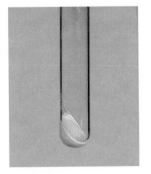

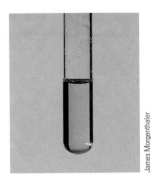

Manganese(II) sulfide, MnS, is salmon-colored. MnS dissolves in 6 M HCl. The resulting solution of $MnCl_2$ is pale pink.

Copper(II) sulfide, CuS, is black. As CuS dissolves in 6 M HNO_3, some NO is oxidized to brown NO_2 by O_2 in the air. The resulting solution of $Cu(NO_3)_2$ is blue.

1215

EXAMPLE 36-1   Complex Ion Formation

What are the concentrations of hydrated Cu^{2+}, NH_3, and $[Cu(NH_3)_4]^{2+}$ in a 0.20 M solution of $[Cu(NH_3)_4]SO_4$?

Plan

We write the equation for the *complete dissociation* of the soluble compound that contains the complex ion. This gives us the concentration of the complex ion $[Cu(NH_3)_4]^{2+}$. Then we write the equation for the dissociation of the complex ion, represent the equilibrium concentrations algebraically, and substitute them into the equilibrium constant expression.

Solution

The soluble deep blue complex salt $[Cu(NH_3)_4]SO_4$ dissociates completely to produce tetraamminecopper(II) ions and sulfate ions.

$$[Cu(NH_3)_4]SO_4(aq) \xrightarrow{100\%} [Cu(NH_3)_4]^{2+}(aq) + SO_4^{2-}(aq)$$
$$0.20\,M \Longrightarrow \quad\quad 0.20\,M \quad\quad 0.20\,M$$

Some of the $[Cu(NH_3)_4]^{2+}$ ions then dissociate. Let x be the concentration of $[Cu(NH_3)_4]^{2+}$ that dissociates.

$$[Cu(NH_3)_4]^{2+} \Longleftrightarrow Cu^{2+}(aq) + 4NH_3$$
$$(0.20 - x)\,M \quad\quad x\,M \quad\quad 4x\,M$$

$$K_d = \frac{[Cu^{2+}][NH_3]^4}{[[Cu(NH_3)_4]^{2+}]} = 8.5 \times 10^{-13} = \frac{x(4x)^4}{0.20 - x} = \frac{256\,x^5}{0.20}$$

[assume that $(0.20 - x) \approx 0.20$]

$$x^5 = 6.6 \times 10^{-16}$$

Taking the fifth root of both sides of this equation gives $x = 9.2 \times 10^{-4}$.

$$[Cu^{2+}] = x\,M = 9.2 \times 10^{-4}\,M$$
$$[NH_3] = 4x\,M = 3.7 \times 10^{-3}\,M$$
$$[[Cu(NH_3)_4]^{2+}] = (0.20 - x)\,M \approx 0.20\,M$$

$(0.20 - x) \approx 0.20$ is a valid assumption; 9.2×10^{-4} is much less than 0.20.

Copper(II) hydroxide dissolves in an excess of aqueous NH_3 to form the deep blue complex ion $[Cu(NH_3)_4]^{2+}$. This decreases the $[Cu^{2+}]$ so that $[Cu^{2+}][OH^-]^2 < K_{sp}$, and so that the $Cu(OH)_2$ dissolves.

$$Cu(OH)_2(s) \Longleftrightarrow Cu^{2+}(aq) \quad\quad + 2OH^-(aq)$$
$$\underline{Cu^{2+}(aq) + 4NH_3(aq) \Longleftrightarrow [Cu(NH_3)_4]^{2+}(aq)}$$
overall rxn: $$Cu(OH)_2(s) + 4NH_3(aq) \Longleftrightarrow [Cu(NH_3)_4]^{2+}(aq) + 2OH^-(aq)$$

On the other hand, zinc hydroxide is amphoteric (Section 10-6 in the textbook). This tells us that solid $Zn(OH)_2$ dissolves in excess NaOH solution to form the complex ion $[Zn(OH)_4]^{2-}$.

$$Zn(OH)_2(s) + 2OH^-(aq) \Longleftrightarrow [Zn(OH)_4]^{2-}(aq)$$

EXAMPLE 36-2   Complex Ion Equilibria

Some solid $Zn(OH)_2$ is suspended in a saturated solution of $Zn(OH)_2$. A solution of sodium hydroxide is added until all the $Zn(OH)_2$ just dissolves. The pH of the solution is 11.80. What

---

As before, the outer brackets mean molar concentrations. The inner brackets are part of the formula of the complex ion.

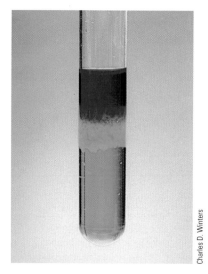

Charles D. Winters

Concentrated aqueous NH_3 was added *slowly* to a solution of copper(II) sulfate, $CuSO_4$. Unreacted blue copper(II) sulfate solution remains in the bottom part of the test tube. The light blue precipitate in the middle is copper(II) hydroxide, $Cu(OH)_2$. The top layer contains deep blue $[Cu(NH_3)_4]^{2+}$ ions that were formed as some $Cu(OH)_2$ dissolved in excess aqueous NH_3.

are the concentrations of Zn^{2+} and $[Zn(OH)_4]^{2-}$ ions in the solution? K_{sp} for $Zn(OH)_2 = 4.5 \times 10^{-17}$, and K_d for $[Zn(OH)_4]^{2-} = 3.5 \times 10^{-16}$.

Plan

We write the appropriate chemical equations and equilibrium constants for the two reversible reactions. We see that $[OH^-]$ appears in both of these equilibrium constant expressions. We are given pH, from which we can calculate $[OH^-]$. Once we know $[OH^-]$, we can use K_{sp} for $Zn(OH)_2$ to calculate $[Zn^{2+}]$. Then, knowing both $[OH^-]$ and $[Zn^{2+}]$, we can solve for $[[Zn(OH)_4]^{2-}]$.

Solution

The important equilibria and their equilibrium constant expressions are

$$Zn(OH)_2(s) \rightleftharpoons Zn^{2+}(aq) + 2OH^-(aq) \qquad K_{sp} = [Zn^{2+}][OH^-]^2 = 4.5 \times 10^{-17}$$

$$[Zn(OH)_4]^{2-} \rightleftharpoons Zn^{2+}(aq) + 4OH^-(aq) \qquad K_d = \frac{[Zn^{2+}][OH^-]^4}{[[Zn(OH)_4]^{2-}]} = 3.5 \times 10^{-16}$$

We know the pH, and so we can calculate $[OH^-]$ from pH + pOH = 14.00.

$$pOH = 14.00 - pH = 14.00 - 11.80 = 2.20$$

$$[OH^-] = 10^{-pOH} = 10^{-2.20} = 6.3 \times 10^{-3}\ M = [OH^-]$$

We can use the solubility product expression for $Zn(OH)_2$ to calculate $[Zn^{2+}]$.

$$K_{sp} = [Zn^{2+}][OH^-]^2 = 4.5 \times 10^{-17}$$

$$[Zn^{2+}] = \frac{K_{sp}}{[OH^-]^2} = \frac{4.5 \times 10^{-17}}{(6.3 \times 10^{-3})^2} = \boxed{1.1 \times 10^{-12}\ M\ Zn^{2+}}$$

Both equilibria are established in the same solution, and so the same $[Zn^{2+}]$ and $[OH^-]$ also satisfy the complex ion dissociation equilibrium.

$$K_d = \frac{[Zn^{2+}][OH^-]^4}{[[Zn(OH)_4]^{2-}]} = 3.5 \times 10^{-16}$$

$$[[Zn(OH)_4]^{2-}] = \frac{[Zn^{2+}][OH^-]^4}{3.5 \times 10^{-16}} = \frac{(1.1 \times 10^{-12})(6.3 \times 10^{-3})^4}{3.5 \times 10^{-16}}$$

$$\boxed{[[Zn(OH)_4]^{2-}] = 5.0 \times 10^{-6}\ M}$$

Thus, we see that we are able to shift equilibria [in this case, dissolve $Zn(OH)_2$] by taking advantage of complex ion formation.

36-2 EQUILIBRIA INVOLVING COMPLEX IONS

Dissolving precipitates by complex ion formation is used at several points in qualitative analysis. *By convention, equilibrium constants for complex ions are usually written as dissociation constants.* We have demonstrated that dissociation constants for complex ions can be treated like other equilibrium constants. The dissociation equation and equilibrium constant expression for the diamminesilver ion are

$$[Ag(NH_3)_2]^+ \rightleftharpoons Ag^+ + 2NH_3 \qquad K_d = \frac{[Ag^+][NH_3]^2}{[[Ag(NH_3)_2]^+]} = 6.3 \times 10^{-8}$$

These equilibrium constants are sometimes represented as formation constants.

$$Ag^+ + 2NH_3 \rightleftharpoons [Ag(NH_3)_2]^+$$

$$K_f = \frac{[[Ag(NH_3)_2]^+]}{[Ag^+][NH_3]^2} = \frac{1}{K_d} = \frac{1}{6.3 \times 10^{-8}} = 1.6 \times 10^7$$

The chlorides of silver and mercury(I) ions, AgCl and Hg_2Cl_2, can be separated by dissolving AgCl in an excess of aqueous ammonia. This is a typical example of the dissolution of a precipitate by forming a soluble complex compound.

$$\left.\begin{array}{l} \underline{AgCl} \\ \underline{Hg_2Cl_2} \end{array}\right\} \xrightarrow[\text{aq. NH}_3]{\text{excess}} \begin{array}{l} [Ag(NH_3)_2]^+ + Cl^- \\ \underline{HgNH_2Cl} + Hg(\ell) \end{array}$$

Example 36-3 illustrates the quantitative aspects of such equilibria in a somewhat simplified format.

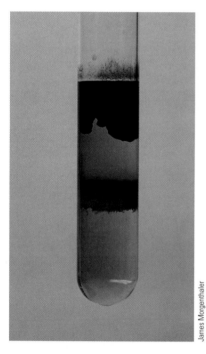

Some equilibria involving Co^{2+} ions.

1. Co^{2+} ions, in the pink solution (*bottom*), react with aqueous NH_3 to form $Co(OH)_2$.
$K_{sp} = 2.5 \times 10^{-16}$.
2. $Co(OH)_2$, the blue-gray solid (*next to bottom layer*), reacts with excess aqueous NH_3 to form $[Co(NH_3)_6]^{2+}$ ions (tan solution).
$K_d = 1.3 \times 10^{-5}$.
3. $[Co(NH_3)_6]^{2+}$ ions, in the tan solution, react with S^{2-} ions to form CoS, the black solid (*top*).
$K_{sp} = 5.9 \times 10^{-21}$.

> ## EXAMPLE 36-3   *Dissolution of a Precipitate*
>
> Calculate the minimum number of moles of gaseous NH_3 required to dissolve 0.020 mole of AgCl in enough water to give 1 liter of solution.

Solution

The dissolution of AgCl in aqueous NH_3 may be represented as

$$\begin{array}{ccccc} AgCl(s) & + & 2NH_3 & \longrightarrow & [Ag(NH_3)_2]^+ & + & Cl^- \\ 0.020 \text{ mol} & & 2(0.020 \text{ mol}) & \Longrightarrow & 0.020 \text{ mol} & & 0.020 \text{ mol} \end{array}$$

This equation tells us that the dissolution of 0.020 mole of AgCl *requires* 2(0.020) mole of NH_3 and *produces* 0.020 mole of $[Ag(NH_3)_2]^+$ and 0.020 mole of Cl^-. Thus, the *stoichiometry of the reaction requires 0.040 mole of NH_3*. In addition, we must calculate the number of moles of NH_3 in excess of this amount that is *required to satisfy the dissociation constant for the complex ion,* $[Ag(NH_3)_2]^+$.

Two equilibria are involved in the dissolution of AgCl in aqueous NH_3. Therefore, both equilibrium constants must be satisfied. They are

$$AgCl(s) \rightleftharpoons Ag^+ + Cl^- \qquad K_{sp} = [Ag^+][Cl^-] = 1.8 \times 10^{-10}$$

$$[Ag(NH_3)_2]^+ \rightleftharpoons Ag^+ + 2NH_3 \qquad K_d = \frac{[Ag^+][NH_3]^2}{[[Ag(NH_3)_2]^+]} = 6.3 \times 10^{-8}$$

Because Ag^+ ion is the species common to both chemical equations and both equilibrium constants, it provides a "connection" between the two equilibria. The equation for the dissolution of AgCl, a reaction that goes to completion, tells us that the solution contains 0.020 M Cl^-. Therefore, we can solve the solubility product expression of AgCl for $[Ag^+]$, which gives us a start on the problem.

$$[Ag^+] = \frac{1.8 \times 10^{-10}}{[Cl^-]} = \frac{1.8 \times 10^{-10}}{0.020} = 9.0 \times 10^{-9} \, M$$

This $[Ag^+]$ is substituted into the dissociation constant expression for $[Ag(NH_3)_2]^+$, which allows us to calculate the *equilibrium concentration of NH_3*. Recall that the stoichiometry of the dissolution reaction tells us that $[[Ag(NH_3)_2]^+] = 0.020 \, M$, so $[NH_3]$ is the only unknown in this expression.

$$[NH_3]^2 = \frac{6.3 \times 10^{-8} \, [[Ag(NH_3)_2]^+]}{[Ag^+]} = \frac{(6.3 \times 10^{-8})(0.020)}{9.0 \times 10^{-9}}$$

$$[NH_3]^2 = 0.14 \qquad [NH_3] = 0.37 \, M$$

The volume of the solution is one liter. So the total number of moles of NH_3 is the number of moles required by the stoichiometry of the dissolution reaction (0.040 mole) *plus* the number of moles required to satisfy the dissociation constant for $[Ag(NH_3)_2]^+$. That is, in terms of concentration

$$[NH_3]_{total} = 0.040 \, M + 0.37 \, M = 0.41 \, M \qquad \therefore \quad \boxed{0.41 \text{ mole of } NH_3 \text{ is required}}$$

The dissolution of 0.020 mole of AgCl requires one liter of 0.41 M aqueous NH_3. The equilibrium concentration of NH_3 is 0.37/0.04 = 9.3 times greater than the stoichiometric amount of NH_3.

K_d is fairly large. Therefore, a large excess of NH_3 is required.

Silver cyanide, AgCN, $K_{sp} = 1.2 \times 10^{-16}$, is much less soluble than silver chloride, $K_{sp} = 1.8 \times 10^{-10}$. However, silver cyanide is readily soluble in solutions of soluble cyanides such as potassium cyanide, KCN, because the $[Ag(CN)_2]^-$ ion is so stable ($K_d = 1.8 \times 10^{-19}$).

EXAMPLE 36-4   *Dissolution of a Precipitate*

Calculate the molarity of a KCN solution, one liter of which will just dissolve 0.020 mole of AgCN.

Solution

Again, we have two equilibria and two equilibrium constants to consider.

$$AgCN(s) \rightleftharpoons Ag^+ + CN^- \qquad K_{sp} = [Ag^+][CN^-] = 1.2 \times 10^{-16}$$

$$[Ag(CN)_2]^- \rightleftharpoons Ag^+ + 2CN^- \qquad K_d = \frac{[Ag^+][CN^-]^2}{[[Ag(CN)_2]^-]} = 1.8 \times 10^{-19}$$

The equation for the dissolution of 0.020 mole of AgCN in KCN solution is

$$\begin{array}{ccc} AgCN(s) + & CN^- & \longrightarrow [Ag(CN)_2]^- \\ 0.020 \text{ mol} & 0.020 \text{ mol} \Longrightarrow & 0.020 \text{ mol} \end{array}$$

The stoichiometry of this reaction tells us that 0.020 mole of CN^- is *required* to dissolve 0.020 mole of AgCN, and this *produces* 0.020 mole of $[Ag(CN)_2]^-$.

Examination of the two equilibrium constant expressions reveals that both $[Ag^+]$ and $[CN^-]$ occur in both. Therefore, we solve for one concentration in terms of the other. We solve the solubility product expression for $[Ag^+]$.

$$[Ag^+][CN^-] = 1.2 \times 10^{-16} \qquad or \qquad [Ag^+] = \frac{1.2 \times 10^{-16}}{[CN^-]}$$

Now we substitute this value into the dissociation constant expression. Recall that $[[Ag(CN)_2]^-] = 0.020 \, M$ (from the stoichiometry of the reaction).

$$\frac{[Ag^+][CN^-]^2}{[[Ag(CN)_2]^-]} = 1.8 \times 10^{-19} \qquad \frac{\left(\dfrac{1.2 \times 10^{-16}}{[CN^-]}\right)[CN^-]^2}{(0.020)} = 1.8 \times 10^{-19}$$

This expression contains only one unknown, [CN⁻]. We solve for [CN⁻].

$$[CN^-] = 3.0 \times 10^{-5}\ M$$

The total number of moles of CN^- is the number *required* by the stoichiometry of the dissolution reaction (0.020 mole) *plus* the number required to satisfy the dissociation constant for $[Ag(CN)_2]^-$ (in terms of concentration):

$$[CN^-]_{total} = 0.020\ M + 3.0 \times 10^{-5}\ M$$

> From a practical point of view, only slightly more than 0.020 mole per liter of KCN, the stoichiometric amount, would be required to dissolve 0.020 mole of AgCN.

K_d is very small. Therefore, only a slight excess of CN⁻ is required.

You might like to verify the answer obtained in Example 36-4 by using the equilibrium concentration of CN^- ions just as we used the equilibrium concentration of Cl^- ions in Example 36-3.

Calculations on other complex species are similar to these examples. However, many complex ions contain three, four, or six ligands (Appendix I), and therefore calculations on these involve higher order equations (Appendix A).

36-3 HYDROGEN SULFIDE EQUILIBRIA

Many equilibria in qualitative analysis involve hydrogen sulfide. In Section 18-5 in the textbook, we described the stepwise ionization of polyprotic acids. Let us consider in detail the equilibria in saturated aqueous solutions of H_2S. Hydrosulfuric acid, H_2S, is a *very weak* diprotic acid. Saturated H_2S solutions are 0.10 M.

EXAMPLE 36-5 *Solutions of Weak Polyprotic Acids*

Calculate the concentrations of the species in 0.10 M H_2S solution. $K_{a1} = 1.0 \times 10^{-7}$ and $K_{a2} = 1.0 \times 10^{-19}$.

Plan

We use the same kind of logic as in Example 18-16 in *General Chemistry* by Whitten, Davis, Peck, and Stanley.

Solution

Let x = mol/L of H_2S that ionize in the first step.

Because H₂S contains two acidic hydrogens per formula unit, we show its ionization in two steps. For the first step, write the appropriate ionization equation with its K_a expression and value. Then, represent the equilibrium concentrations from the first ionization step, and substitute into the K_{a1} expression. Repeat the process for the second ionization step.

$$\begin{array}{ccccccc} H_2S & + & H_2O & \rightleftharpoons & H_3O^+ & + & HS^- \\ (0.10-x)\,M & & & & x\,M & & x\,M \end{array}$$

$$K_{a1} = \frac{[H_3O^+][HS^-]}{[H_2S]} = \frac{(x)(x)}{(0.10-x)} = 1.0 \times 10^{-7}$$

Solving with the usual simplifying assumption gives $x = 1.0 \times 10^{-4}$.

$$[H_2S] = (0.10-x)\,M = \boxed{0.10\ M}$$

$$[HS^-] = [H_3O^+] = x\,M = \boxed{1.0 \times 10^{-4}\ M} \qquad \text{first step}$$

The second step involves ionization of the anion produced in the first step. We represent equilibrium concentrations (where y = mol/L of HS^- that ionizes) as follows.

$$\text{HS}^- \quad + \text{H}_2\text{O} \rightleftharpoons \quad \text{H}_3\text{O}^+ \quad + \text{S}^{2-}$$

$$(1.0 \times 10^{-4} - y)\,M \qquad\qquad (1.0 \times 10^{-4} + y)\,M \quad y\,M$$

from 1st step   from 2nd step

Substitution into K_{a2} gives

$$K_{a2} = \frac{[\text{H}_3\text{O}^+][\text{S}^{2-}]}{[\text{HS}^-]} = \frac{(1.0 \times 10^{-4} + y)(y)}{(1.0 \times 10^{-4} - y)} = 1.0 \times 10^{-19}$$

Assume that $(1.0 \times 10^{-4} + y) \approx 1.0 \times 10^{-4}$ and $(1.0 \times 10^{-4} - y) \approx 1.0 \times 10^{-4}$.

$$\frac{(1.0 \times 10^{-4})y}{1.0 \times 10^{-4}} = 1.0 \times 10^{-19} \qquad y = \boxed{1.0 \times 10^{-19}\,M = [\text{HS}^-]} = [\text{H}_3\text{O}^+]_{2\text{nd}}$$

Our assumption is valid. Very little HS^- ionizes ($1.0 \times 10^{-19}\,M$).

$$[\text{S}^{2-}] = y = \boxed{1.0 \times 10^{-19}\,M}$$

$$[\text{HS}^-] = (1.0 \times 10^{-4} - y) = (1.0 \times 10^{-4}) - (1.0 \times 10^{-19}) \approx \boxed{1.0 \times 10^{-4}\,M}$$

$$[\text{H}_3\text{O}^+] = (1.0 \times 10^{-4} + y) = (1.0 \times 10^{-4}) + (1.0 \times 10^{-19}) \approx \boxed{1.0 \times 10^{-4}\,M}$$

We have already calculated the $\boxed{\text{concentration of nonionized } \text{H}_2\text{S}\ (\approx 0.10\,M).}$

These concentrations (Example 36-5) tell us that saturated solutions of H_2S ($0.10\,M$) are only slightly acidic. They contain very low concentrations of S^{2-} ions. Recall that for solutions containing only weak polyprotic acids in reasonable concentrations, $[\text{anion}^{2-}] = K_{a2}$.

H_2S is a very weak acid. Many experiments have shown that in solutions containing H_2S *and* a strong acid, such as HCl, the acidity is determined by the strong acid. We can derive a simple relationship between $[\text{H}^+]$ and $[\text{S}^{2-}]$ in such solutions. We multiply K_{a1} and K_{a2}, the ionization constants for H_2S,

$$K_{a1}K_{a2} = \frac{[\text{H}^+][\text{HS}^-]}{[\text{H}_2\text{S}]} \times \frac{[\text{H}^+][\text{S}^{2-}]}{[\text{HS}^-]} = (1.0 \times 10^{-7})(1.0 \times 10^{-19})$$

to obtain

$$\frac{[\text{H}^+]^2[\text{S}^{2-}]}{[\text{H}_2\text{S}]} = 1.0 \times 10^{-26}$$

If we restrict our discussions to *saturated solutions* of H_2S ($0.10\,M$), we can simplify the relationship further.

$$[\text{H}^+]^2[\text{S}^{2-}] = (1.0 \times 10^{-26})[\text{H}_2\text{S}] = (1.0 \times 10^{-26})(0.10)$$

$$[\text{H}^+]^2[\text{S}^{2-}] = 1.0 \times 10^{-27} \qquad \text{valid for saturated } \text{H}_2\text{S} \text{ solutions that contain a strong acid}$$

This is a very useful relationship for *saturated* H_2S solutions. It tells us that the concentration of sulfide ions *varies inversely with the square of the concentration of hydrogen (hydronium) ions.*

> ### EXAMPLE 36-6   *The Common Ion Effect in H$_2$S Solutions*
>
> Calculate the concentrations of H$^+$ and S^{2-} ions in a saturated (0.10 M) H$_2$S solution that is also 0.30 M in HCl, the concentrations used to precipitate Group I sulfides.
>
> **Solution**
>
> Example 36-5 showed that the maximum [H$^+$] in saturated H$_2$S is 1.0×10^{-4} M. Because [H$^+$] from 0.30 M HCl, a strong acid, is 0.30 M,
>
> $$[H^+]_{HCl} \gg [H^+]_{H_2S}$$
>
> $$0.30\ M \gg 0.00010\ M \qquad \therefore \quad \boxed{[H^+] = 0.30\ M}$$
>
> Substitution of [H$^+$] = 0.30 M into the relationship we just derived for saturated H$_2$S solutions gives the value for [S^{2-}].
>
> $$[S^{2-}] = \frac{1.0 \times 10^{-27}}{[H^+]^2} = \frac{1.0 \times 10^{-27}}{(0.30)^2} = \boxed{1.1 \times 10^{-26}\ M}$$

Note that [S^{2-}] = 1.1×10^{-26} M in saturated H$_2$S that is also 0.30 M in HCl. Example 36-5 showed that [S^{2-}] = 1.0×10^{-19} M is saturated H$_2$S. Therefore, the presence of the strong acid has decreased [S^{2-}] by the factor $1.0 \times 10^{-19}/1.1 \times 10^{-26}$, or 9.1×10^6. Stated differently, the concentration of S^{2-} is more than 9 million times greater in saturated H$_2$S than in saturated H$_2$S that is *also* 0.30 M in HCl (or any other strong monoprotic acid).

> ### EXAMPLE 36-7   *The Common Ion Effect in H$_2$S Solutions*
>
> Calculate [S^{2-}] in saturated H$_2$S that is also 0.60 M in HCl.
>
> **Solution**
>
> As in Example 36-6, we substitute [H$^+$], which is 0.60 M in this example, into our useful relationship for saturated H$_2$S solutions and calculate [S^{2-}].
>
> $$[H^+]^2[S^{2-}] = 1.0 \times 10^{-27}$$
>
> $$[S^{2-}] = \frac{1.0 \times 10^{-27}}{[H^+]^2} = \frac{1.0 \times 10^{-27}}{(0.60)^2} = \boxed{2.8 \times 10^{-27}\ M}$$

We see that this [S^{2-}] is smaller than that in Example 36-6 because the concentration of HCl is twice as great. The ratio of S^{2-} ion concentrations in the two solutions is $1.1 \times 10^{-26}/2.8 \times 10^{-27} = 4$ (within roundoff-error range). Doubling [H$^+$] decreases [S^{2-}] by a factor of four. Halving [H$^+$] would increase [S^{2-}] by a factor of four, a typical inverse square relationship.

36-4   PRECIPITATION OF METAL SULFIDES

You may find it helpful to review Section 31-5 carefully.

In discussing the precipitation of the Group I sulfides, we indicated that unstable intermediate compounds containing HS$^-$ ions are probably formed. However, because equilibrium depends on only the reactants and products of chemical reactions, we can describe reactions in which insoluble sulfides precipitate.

EXAMPLE 36-8   Simultaneous Equilibria in Precipitation Reactions

Will copper(II) sulfide precipitate in a solution that is 0.10 M in copper(II) nitrate, 0.30 M in HCl, and saturated with H_2S?

Solution

In Example 36-6 we demonstrated that $[S^{2-}] = 1.1 \times 10^{-26}$ M in such a solution. Because $Cu(NO_3)_2$ is a soluble ionic compound, we know that $[Cu^{2+}] = 0.10$ M in 0.10 M $Cu(NO_3)_2$. We calculate Q_{sp} and compare it with K_{sp} for CuS.

$$K_{sp} = [Cu^{2+}][S^{2-}] = 8.7 \times 10^{-36}$$

$$Q_{sp} = (0.10)(1.1 \times 10^{-26}) = 1.1 \times 10^{-27}$$

Because $Q_{sp} \gg K_{sp}$ for CuS, precipitation occurs until $Q_{sp} = K_{sp}$.

We can also calculate the maximum concentration of an ion that can exist in a particular solution (see also Section 31-5).

EXAMPLE 36-9   Simultaneous Equilibria in Precipitation Reactions

What is the maximum concentration of Cu^{2+} ions that can exist (without forming a precipitate) in a saturated H_2S solution that is also 0.30 M in HCl?

Solution

Because we know that $[S^{2-}] = 1.1 \times 10^{-26}$ M in this solution (from Example 36-6), we can use K_{sp} for CuS to determine the $[Cu^{2+}]$.

$$[Cu^{2+}][S^{2-}] = 8.7 \times 10^{-36}$$

$$[Cu^{2+}] = \frac{8.7 \times 10^{-36}}{[S^{2-}]} = \frac{8.7 \times 10^{-36}}{1.1 \times 10^{-26}} = 7.9 \times 10^{-10}$$

This is only 5.0×10^{-8} g Cu^{2+}/L, and so we conclude that only a tiny trace of Cu^{2+} ions can exist in such solutions.

Note that the statement of Example 36-9 asked for the maximum $[Cu^{2+}]$ that could exist in a particular solution, *not* the $[Cu^{2+}]$ remaining after precipitation. We shall deal with that subject presently.

The equations for the reactions by which the Groups I and II sulfides are precipitated show that many of these reactions produce H^+ ions. As these reactions occur, the acidity of the solution increases (Example 36-10).

EXAMPLE 36-10   Simultaneous Equilibria in Precipitation Reactions

A solution is initially 0.10 M in $Cu(NO_3)_2$ and 0.30 M in HCl. Then sufficient H_2S is added so that the solution is saturated with H_2S. Calculate the concentration of Cu^{2+} ions that remains in the solution when precipitation is as complete as possible.

Solution

The precipitation reaction produces hydrogen ions.

$$Cu^{2+} + H_2S \longrightarrow CuS(s) + 2H^+$$
$$0.10\ M \quad \text{excess} \Longrightarrow \quad\quad\quad 0.20\ M$$

The very small K_{sp} for CuS, 8.7×10^{-36}, tells us that CuS is quite insoluble. Therefore, we *assume* that precipitation is essentially complete. We started with 0.10 *M* Cu(NO$_3$)$_2$, a soluble ionic compound, so the reaction produces 0.20 *M* H$^+$. The total [H$^+$] is the concentration produced by the reaction *plus* the 0.30 *M* furnished by 0.30 *M* HCl.

$$[H^+]_{total} = [H^+]_{reaction} + [H^+]_{HCl} = 0.20\ M + 0.30\ M = 0.50\ M$$

We use the familiar relationship to calculate [S^{2-}] now that [H$^+$]$_{total}$ is known.

$$[H^+]^2[S^{2-}] = 1.0 \times 10^{-27}$$

$$[S^{2-}] = \frac{1.0 \times 10^{-27}}{[H^+]^2} = \frac{1.0 \times 10^{-27}}{(0.50)^2} = 4.0 \times 10^{-27}$$

We know [S^{2-}] after precipitation has occurred, so we use K_{sp} to calculate [Cu^{2+}].

$$[Cu^{2+}][S^{2-}] = 8.7 \times 10^{-36}$$

$$[Cu^{2+}] = \frac{8.7 \times 10^{-36}}{[S^{2-}]} = \frac{8.7 \times 10^{-36}}{4.0 \times 10^{-27}} = \boxed{2.2 \times 10^{-9}\ M}$$

The concentration of copper(II) ions remaining in solution is only 2.2×10^{-9} *M*. Our assumption that precipitation would be essentially complete is valid.

The acidity or basicity of the reaction medium is an extremely important factor in many chemical reactions. By carefully controlling the pH of a solution, it is possible to precipitate some ions nearly completely while others are left in solution. Consider a solution that contains copper(II) nitrate, Cu(NO$_3$)$_2$; cadmium nitrate, Cd(NO$_3$)$_2$; and zinc nitrate, Zn(NO$_3$)$_2$, all soluble ionic compounds. The three metal ions, Cu^{2+}, Cd^{2+}, and Zn^{2+}, all form insoluble sulfides. Their solubility products indicate significant differences in the solubilities of these "insoluble" compounds. Clearly, CuS is the least soluble and ZnS is the most soluble.

Compound	K_{sp}	Difference (as a ratio)
ZnS	$[Zn^{2+}][S^{2-}] = 1.1 \times 10^{-21}$	$\left.\vphantom{\begin{array}{c}a\\b\end{array}}\right\}\ 3.1 \times 10^7$
CdS	$[Cd^{2+}][S^{2-}] = 3.6 \times 10^{-29}$	$\left.\vphantom{\begin{array}{c}a\\b\end{array}}\right\}\ 4.1 \times 10^6$
CuS	$[Cu^{2+}][S^{2-}] = 8.7 \times 10^{-36}$	

If a solution is made 0.30 *M* in HCl and 0.010 *M* each in Cu(NO$_3$)$_2$, Cd(NO$_3$)$_2$, and Zn(NO$_3$)$_2$, and then saturated with H$_2$S (0.10 *M*), will ZnS precipitate? To answer this question, we must determine whether the solubility product for ZnS is exceeded, and so we must calculate both [Zn^{2+}] and [S^{2-}]. From Example 36-6, we know that [S^{2-}] = 1.1×10^{-26} *M* in 0.30 *M* HCl saturated with H$_2$S.

The K_{sp} expressions for CuS, CdS, and ZnS are all of the same form, [M^{2+}][S^{2-}] = K_{sp}. Because [S^{2-}] is known, we can determine whether any or all of the insoluble sulfides will precipitate from the solution under discussion.

$$K_{sp} = [Zn^{2+}][S^{2-}] = 1.1 \times 10^{-21}$$

In 0.010 M $Zn(NO_3)_2$, we have $[Zn^{2+}] = 0.010\ M$, so

$$Q_{sp} = [Zn^{2+}][S^{2-}] = (0.010)(1.1 \times 10^{-26}) = 1.1 \times 10^{-28}$$
$$Q_{sp} = 1.1 \times 10^{-28} < 1.1 \times 10^{-21} = K_{sp}$$

This calculation tells us that ZnS does *not* precipitate from the solution.

Now let us see if CdS and CuS do precipitate from this solution. For cadmium sulfide we know that

$$K_{sp} = [Cd^{2+}][S^{2-}] = 3.6 \times 10^{-29}$$

In this solution,

$$[Cd^{2+}] = 0.010\ M \qquad [S^{2-}] = 1.1 \times 10^{-26}$$
$$Q_{sp} = [Cd^{2+}][S^{2-}] = (0.010)(1.1 \times 10^{-26}) = 1.1 \times 10^{-28}$$
$$Q_{sp} = 1.1 \times 10^{-28} > 3.6 \times 10^{-29} = K_{sp}$$

This tells us that CdS *does* precipitate from the solution. Because CuS is less soluble than CdS, clearly it will precipitate also.

We have described the reactions that occur in a solution that is

0.30 M in HCl,

0.010 M each in $Cu(NO_3)_2$, $Cd(NO_3)_2$ and $Zn(NO_3)_2$, and saturated with H_2S

Copper(II) sulfide and cadmium sulfide precipitate, but zinc sulfide does not.

The solubility behavior of insoluble sulfides, carbonates, phosphates, and arsenates is quite complex due to extensive hydrolysis of these anions. The situation is further complicated when certain cations also undergo significant hydrolysis. Detailed calculations on the effect of hydrolysis on the solubilities of such compounds are beyond the scope of this text.

Exercises

Dissolution of Precipitates and Complex Ion Formation

1. Explain, by writing appropriate equations, how the following insoluble compounds can be dissolved by the addition of a solution of nitric acid. (Carbonates dissolve in strong acids to form carbon dioxide, which is evolved as a gas, and water.) What is the "driving force" for each reaction? (a) $Cu(OH)_2$; (b) $Al(OH)_3$; (c) $MnCO_3$.

2. Explain, by writing equations, how the following insoluble compounds can be dissolved by the addition of a solution of ammonium nitrate or ammonium chloride. (a) $Mg(OH)_2$; (b) $Mn(OH)_2$; (c) $Ni(OH)_2$.

3. The following insoluble sulfides can be dissolved in 3 M hydrochloric acid. Explain how this is possible and write the appropriate equations. (a) MnS; (b) FeS.

4. The following sulfides are less soluble than those listed in Exercise 3 and can be dissolved in hot 6 M nitric acid, an oxidizing acid. Explain how, and write the appropriate balanced equations. (a) PbS; (b) CuS; (c) Bi_2S_3.

5. Why would MnS be expected to be more soluble in 0.10 M HCl solution than in water? Would the same be true for $Mn(NO_3)_2$?

6. How can most water-insoluble metal hydroxides be dissolved? Write a chemical equation for the dissolution of $Fe(OH)_3$ in this way.

7. How does the presence of excess H_3O^+ aid in the dissolution of slightly soluble metal carbonates? Write the chemical equation for the dissolution of $MnCO_3$.

8. Find the concentration of Au^{3+} in a solution in which the other equilibrium concentrations are $[Cl^-] = 0.10\ M$ and $[AuCl_4^-] = 0.20\ M$. K_d for $AuCl_4^- = 7.0 \times 10^{-26}$.

Complex Ion Equilibria

9. (a) What is the relationship between the dissociation constant and the formation constant for a complex ion? (b) Illustrate the relationship for $[AgCl_2]^-$, $[Cd(NH_3)_4]^{2+}$, and $[AlF_6]^{3-}$. Appendix I will be helpful.

10. Write equations and equilibrium constant expressions for the equilibria involved in, as well as the overall equation for, the dissolution of (a) AgCl in excess aqueous NH_3, (b) $Co(OH)_2$ in excess aqueous NH_3, (c) $PbCl_2$ in excess HCl, and (d) AgCl in excess $Na_2S_2O_3$.

11. Calculate the concentrations of complex ion, metal ion, and ammonia in the following solutions. The complex ions are enclosed in brackets. All these compounds are soluble and ionic.
    (a) 0.100 M [Ag(NH_3)_2]Cl
    (b) 0.089 M [Co(NH_3)_6]SO_4
    (c) 0.114 M [Co(NH_3)_6]_2(SO_4)_3

*12. What is the concentration of Ag^+ in a solution that is 0.10 M in KSCN and 0.10 M in [Ag(SCN)_4]^{3-}? Will Ag_2SO_4 precipitate if [SO_4^{2-}] = 0.10 M? K_d for [Ag(SCN)_4]^{3-} = 2.1×10^{-10}.

13. Calculate the minimum number of moles of gaseous NH_3 required to dissolve the indicated number of moles of AgCl in enough water to give 1.0 L of solution: (a) 0.15 mol; (b) 0.015 mol; (c) 1.5×10^{-3} mol.

14. Calculate the minimum number of moles of the indicated substance required to dissolve 0.025 mol of the indicated salt in enough water to give 1.0 L of solution. (a) AgBr in excess HBr; (b) AgCN in excess NaCN; (c) AgBr in excess $Na_2S_2O_3$; (d) $Cd(CN)_2$ in excess NaCN; (e) $Cd(CN)_2$ in excess NaCl.

Hydrogen Sulfide Equilibria

15. Calculate the concentration of each species in saturated H_2S solution (0.10 M H_2S).

16. Derive the relationship between [H^+] and [S^{2-}] in saturated H_2S solutions that also contain a strong acid such as HCl.

17. What are the concentrations of H^+ and S^{2-} ions in saturated H_2S solutions (0.10 M H_2S) that are also (a) 0.060 M in HCl and (b) 0.030 M in HCl? What is an inverse square relationship?

18. If a saturated H_2S solution is also 0.15 M in HCl and 0.10 M in each of the following metal nitrates, which metal ions would precipitate as insoluble sulfides? Justify your answers by appropriate calculations. (a) $Hg(NO_3)_2$; (b) $Cd(NO_3)_2$; (c) $Zn(NO_3)_2$; (d) $Mn(NO_3)_2$.

*19. (a) What is the maximum concentration of each metal ion listed in Exercise 18 that can exist in a solution that is saturated with H_2S and also 0.040 M in HCl? What does the answer you obtained for Mn^{2+} indicate? (b) How many grams of each metal per liter of solution is this? Does the answer you obtain for Mn^{2+} "make sense"? Why?

*20. Each of the following solutions is 0.025 M in HCl, and sufficient H_2S is added so that the solutions are saturated with H_2S when precipitation is as complete as possible. What concentration(s) of metal ion(s) remain(s) in each solution? (a) 0.10 M $Hg(NO_3)_2$, (b) 0.10 M $Cd(NO_3)_2$, (c) 0.10 M $Hg(NO_3)_2$ and 0.10 M $Cd(NO_3)_2$.

21. Calculate the percentage of each metal ion that is precipitated in (a) and (b) in Exercise 20.

Mixed Exercises

*22. A solution is 0.010 M with respect to Cd^{2+} ions and also 0.010 M with respect to Pd^{2+} ions. Solid KBr is added to the solution (assume negligible change in volume) until [Br^-] = 1.0 M. Use the overall dissociation constants of the following complexes to calculate the concentrations of Cd^{2+} ions and Pd^{2+} ions once equilibrium is reached. K_d for [CdBr_4]^{2-} = 2.0×10^{-4} and for [PdBr_4]^{2-} = 7.7×10^{-14}.

23. What concentration of Ag^+ ions remains in a solution that originally contained 0.10 M Ag^+ ions and 1.3 M NH_3?

*24. 0.010 mol of solid $Zn(OH)_2$ is suspended in a saturated $Zn(OH)_2$ solution. Some 6.0 M NaOH solution is added and the mixture is stirred vigorously. The volume of the solution is now 400 mL, and its pH is 13.15. (a) Does all the $Zn(OH)_2$ dissolve? (b) What is/was the minimum [OH^-] necessary to dissolve $Zn(OH)_2$ completely?

*25. A concentrated, strong acid is added to a solid mixture of 0.010-mol samples of $Fe(OH)_2$ and $Cu(OH)_2$ placed in 1.0 L of water. At what values of pH will the dissolution of each hydroxide be complete? (Assume negligible volume change.)

26. A solution is 0.010 M in I^- ions and 0.010 M in Br^- ions. Ag^+ ions are introduced to the solution by the addition of solid $AgNO_3$. Determine (a) which compound will precipitate first, AgI or AgBr, and (b) the percentage of the halide ion in the first precipitate that is removed from solution before the precipitation of the second compound begins.

SOME MATHEMATICAL OPERATIONS

In chemistry we frequently use very large or very small numbers. Such numbers are conveniently expressed in *scientific*, or *exponential*, *notation*.

A-1 SCIENTIFIC NOTATION

In scientific notation, a number is expressed as the *product of two numbers*. By convention, the first number, called the digit term, is between 1 and 10. The second number, called the *exponential term*, is an integer power of 10. Some examples follow.

$$
\begin{aligned}
10000 &= 1 \times 10^4 & 24327 &= 2.4327 \times 10^4 \\
1000 &= 1 \times 10^3 & 7958 &= 7.958 \times 10^3 \\
100 &= 1 \times 10^2 & 594 &= 5.94 \times 10^2 \\
10 &= 1 \times 10^1 & 98 &= 9.8 \times 10^1 \\
1 &= 1 \times 10^0 \\
1/10 = 0.1 &= 1 \times 10^{-1} & 0.32 &= 3.2 \times 10^{-1} \\
1/100 = 0.01 &= 1 \times 10^{-2} & 0.067 &= 6.7 \times 10^{-2} \\
1/1000 = 0.001 &= 1 \times 10^{-3} & 0.0049 &= 4.9 \times 10^{-3} \\
1/10000 = 0.0001 &= 1 \times 10^{-4} & 0.00017 &= 1.7 \times 10^{-4}
\end{aligned}
$$

Recall that, by definition, (any base)^0 = 1.

The exponent of 10 is the number of places the decimal point must be shifted to give the number in long form. A *positive exponent* indicates that the decimal point is *shifted right* that number of places. A *negative exponent* indicates that the decimal point is *shifted left*. When numbers are written in *standard scientific notation*, there is one nonzero digit to the left of the decimal point.

$$
7.3 \times 10^3 = 73 \times 10^2 = 730 \times 10^1 = 7300
$$
$$
4.36 \times 10^{-2} = 0.436 \times 10^{-1} = 0.0436
$$
$$
0.00862 = 0.0862 \times 10^{-1} = 0.862 \times 10^{-2} = 8.62 \times 10^{-3}
$$

In scientific notation the digit term indicates the number of significant figures in the number. The exponential term merely locates the decimal point and does not represent significant figures.

Addition and Subtraction

In addition and subtraction all numbers are converted to the same power of 10, and the digit terms are added or subtracted.

$$
(4.21 \times 10^{-3}) + (1.4 \times 10^{-4}) = (4.21 \times 10^{-3}) + (0.14 \times 10^{-3}) = \underline{4.35 \times 10^{-3}}
$$
$$
(8.97 \times 10^4) - (2.31 \times 10^3) = (8.97 \times 10^4) - (0.231 \times 10^4) = \underline{8.74 \times 10^4}
$$

Multiplication

The digit terms are multiplied in the usual way, the exponents are added algebraically, and the product is written with one nonzero digit to the left of the decimal.

Two significant figures in answer.

$$(4.7 \times 10^7)(1.6 \times 10^2) = (4.7)(1.6) \times 10^{7+2} = 7.52 \times 10^9 = \underline{7.5 \times 10^9}$$

Two significant figures in answer.

$$(8.3 \times 10^4)(9.3 \times 10^{-9}) = (8.3)(9.3) \times 10^{4-9} = 77.19 \times 10^{-5} = \underline{7.7 \times 10^{-4}}$$

Division

The digit term of the numerator is divided by the digit term of the denominator, the exponents are subtracted algebraically, and the quotient is written with one nonzero digit to the left of the decimal.

$$\frac{8.4 \times 10^7}{2.0 \times 10^3} = \frac{8.4}{2.0} \times 10^{7-3} = \underline{4.2 \times 10^4}$$

Three significant figures in answer.

$$\frac{3.81 \times 10^9}{8.412 \times 10^{-3}} = \frac{3.81}{8.412} \times 10^{[9-(-3)]} = 0.45292 \times 10^{12} = \underline{4.53 \times 10^{11}}$$

Powers of Exponentials

The digit term is raised to the indicated power, and the exponent is multiplied by the number that indicates the power.

$$(1.2 \times 10^3)^2 = (1.2)^2 \times 10^{3 \times 2} = 1.44 \times 10^6 = \underline{1.4 \times 10^6}$$
$$(3.0 \times 10^{-3})^4 = (3.0)^4 \times 10^{-3 \times 4} = 81 \times 10^{-12} = \underline{8.1 \times 10^{-11}}$$

These instructions are applicable to most calculators. If your calculator has other notation, consult your calculator's instruction booklet.

Electronic Calculators   *To square a number:* (1) enter the number and (2) touch the (x^2) button.

$$(7.3)^2 = 53.29 = \underline{53} \qquad \text{(two sig. figs.)}$$

To raise a number y to power x: (1) enter the number; (2) touch the (y^x) button; (3) enter the power; and (4) touch the (=) button.

$$(7.3)^4 = 2839.8241 = \underline{2.8 \times 10^3} \qquad \text{(two sig. figs.)}$$
$$(7.30 \times 10^2)^5 = 2.0730716 \times 10^{14} = \underline{2.07 \times 10^{14}} \qquad \text{(three sig. figs.)}$$

Roots of Exponentials

The exponent must be divisible by the desired root if a calculator is not used. The root of the digit term is extracted in the usual way, and the exponent is divided by the desired root.

$$\sqrt{2.5 \times 10^5} = \sqrt{25 \times 10^4} = \sqrt{25} \times \sqrt{10^4} = \underline{5.0 \times 10^2}$$
$$\sqrt[3]{2.7 \times 10^{-8}} = \sqrt[3]{27 \times 10^{-9}} = \sqrt[3]{27} \times \sqrt[3]{10^{-9}} = \underline{3.0 \times 10^{-3}}$$

Electronic Calculators   *To extract the square root of a number:* (1) enter the number and (2) touch the $(\sqrt{x})$ button.

$$\sqrt{23} = 4.7958315 = \underline{4.8} \qquad \text{(two sig. figs.)}$$

On some models, this function is performed by the $\sqrt[x]{y}$ button.

To extract some other root: (1) enter the number y; (2) touch the (INV) and then the (y^x) button; (3) enter the root to be extracted, x; and (4) touch the (=) button.

A-2 LOGARITHMS

The logarithm of a number is the power to which a base must be raised to obtain the number. Two types of logarithms are frequently used in chemistry: (1) common logarithms (abbreviated log), whose base is 10, and (2) natural logarithms (abbreviated ln), whose base is $e = 2.71828$. The general properties of logarithms are the same no matter what base is used. Many equations in science were derived by the use of calculus, and these often involve natural (base e) logarithms. The relationship between $\log x$ and $\ln x$ is as follows.

$$\ln x = 2.303 \log x$$

$\ln 10 = 2.303$

Finding Logarithms   The common logarithm of a number is the power to which 10 must be raised to obtain the number. The number 10 must be raised to the third power to equal 1000. Therefore, the logarithm of 1000 is 3, written $\log 1000 = 3$. Some examples follow.

Number	Exponential Expression	Logarithm
1000	10^3	3
100	10^2	2
10	10^1	1
1	10^0	0
$1/10 = 0.1$	10^{-1}	-1
$1/100 = 0.01$	10^{-2}	-2
$1/1000 = 0.001$	10^{-3}	-3

To obtain the logarithm of a number other than an integral power of 10, you must use either an electronic calculator or a logarithm table. On most calculators, you do this by (1) entering the number and then (2) pressing the (log) button.

$$\log 7.39 = 0.8686444 = \underline{0.869}$$
$$\log 7.39 \times 10^3 = 3.8686 \quad = \underline{3.869}$$
$$\log 7.39 \times 10^{-3} = -2.1314 \quad = \underline{-2.131}$$

The number to the left of the decimal point in a logarithm is called the *characteristic*, and the number to the right of the decimal point is called the *mantissa*. The characteristic only locates the decimal point of the number, so it is usually not included when counting significant figures. The mantissa has as many significant figures as the number whose log was found.

To obtain the natural logarithm of a number on an electronic calculator, (1) enter the number and (2) press the (ln) or (ln x) button.

$$\ln 4.45 = 1.4929041 = \underline{1.493}$$
$$\ln 1.27 \times 10^3 = 7.1468 \quad = \underline{7.147}$$

Finding Antilogarithms   Sometimes we know the logarithm of a number and must find the number. This is called finding the *antilogarithm* (or *inverse logarithm*). To do this on a calculator, (1) enter the value of the log; (2) press the (INV) button; and (3) press the (log) button.

$$\log x = 6.131; \quad \text{so } x = \text{inverse log of } 6.131 = \underline{1.352 \times 10^6}$$
$$\log x = -1.562; \quad \text{so } x = \text{inverse log of } -1.562 = \underline{2.74 \times 10^{-2}}$$

On some calculators, the inverse log is found as follows:
1. enter the value of the log
2. press the (2ndF) (second function) button
3. press (10^x)

On some calculators, the inverse natural logarithm is found as follows:
1. enter the value of the ln
2. press the (2ndF) (second function) button
3. press (e^x)

To find the inverse natural logarithm, (1) enter the value of the ln; (2) press the (INV) button; and (3) press the (ln) or (ln x) button.

$$\ln x = 3.552; \qquad \text{so } x = \text{inverse ln of } 3.552 = \underline{3.49 \times 10^1}$$

$$\ln x = -1.248; \qquad \text{so } x = \text{inverse ln of } -1.248 = \underline{2.87 \times 10^{-1}}$$

Calculations Involving Logarithms

Because logarithms are exponents, operations involving them follow the same rules as the use of exponents. The following relationships are useful.

$$\log xy = \log x + \log y \qquad \text{or} \qquad \ln xy = \ln x + \ln y$$

$$\log \frac{x}{y} = \log x - \log y \qquad \text{or} \qquad \ln \frac{x}{y} = \ln x - \ln y$$

$$\log x^y = y \log x \qquad \text{or} \qquad \ln x^y = y \ln x$$

$$\log \sqrt[y]{x} = \log x^{1/y} = \frac{1}{y} \log x \qquad \text{or} \qquad \ln \sqrt[y]{x} = \ln x^{1/y} = \frac{1}{y} \ln x$$

A-3   QUADRATIC EQUATIONS

Algebraic expressions of the form

$$ax^2 + bx + c = 0$$

are called **quadratic equations.** Each of the constant terms (a, b, and c) may be either positive or negative. All quadratic equations may be solved by the **quadratic formula.**

$$x = \frac{-b \pm \sqrt{b^2 - 4ac}}{2a}$$

If we wish to solve the quadratic equation $3x^2 - 4x - 8 = 0$, we use $a = 3$, $b = -4$, and $c = -8$. Substitution of these values into the quadratic formula gives

$$x = \frac{-(-4) \pm \sqrt{(-4)^2 - 4(3)(-8)}}{2(3)} = \frac{4 \pm \sqrt{16 + 96}}{6}$$

$$= \frac{4 \pm \sqrt{112}}{6} = \frac{4 \pm 10.6}{6}$$

The two roots of this quadratic equation are

$$\underline{x = 2.4} \qquad \text{and} \qquad \underline{x = -1.1}$$

As you construct and solve quadratic equations based on the observed behavior of matter, you must decide which root has physical significance. Examination of the *equation that defines x* always gives clues about possible values for x. In this way you can tell which is extraneous (has no physical significance). Negative roots are often extraneous.

When you have solved a quadratic equation, you should always check the values you obtained by substitution into the original equation. In the preceding example we obtained $x = 2.4$ and $x = -1.1$. Substitution of these values into the original quadratic equation, $3x^2 - 4x - 8 = 0$, shows that both roots are correct. Such substitutions often do not give a perfect check because some round-off error has been introduced.

ELECTRON CONFIGURATIONS OF THE ATOMS OF THE ELEMENTS

Group	IA	IIA	IIIB		IVB	VB	VIB	VIIB	VIIIB			IB	IIB	IIIA	IVA	VA	VIA	VIIA	VIIIA
	(1)	(2)	(3)		(4)	(5)	(6)	(7)	(8)	(9)	(10)	(11)	(12)	(13)	(14)	(15)	(16)	(17)	(18)
Period 1	1 H **1s**																		1 H / 2 He **1s**
2	3 Li / 4 Be **2s**													5 B	6 C	7 N	8 O	9 F	10 Ne **2p**
3	11 Na / 12 Mg **3s**													13 Al	14 Si	15 P	16 S	17 Cl	18 Ar **3p**
4	19 K	20 Ca	21 Sc **4s**		22 Ti	23 V	24 Cr	25 Mn	26 Fe	27 Co	28 Ni	29 Cu	30 Zn **3d**	31 Ga	32 Ge	33 As	34 Se	35 Br	36 Kr **4p**
5	37 Rb	38 Sr	39 Y **5s**		40 Zr	41 Nb	42 Mo	43 Tc	44 Ru	45 Rh	46 Pd	47 Ag	48 Cd **4d**	49 In	50 Sn	51 Sb	52 Te	53 I	54 Xe **5p**
6	55 Cs	56 Ba	57 La **6s**		72 Hf	73 Ta	74 W	75 Re	76 Os	77 Ir	78 Pt	79 Au	80 Hg **5d**	81 Tl	82 Pb	83 Bi	84 Po	85 At	86 Rn **6p**
7	87 Fr	88 Ra	89 Ac **7s**		104 Rf	105 Db	106 Sg	107 Bh	108 Hs	109 Mt	110	111	112 **6d**						

f-block:

58 Ce	59 Pr	60 Nd	61 Pm	62 Sm	63 Eu	64 Gd	65 Tb	66 Dy	67 Ho	68 Er	69 Tm	70 Yb	71 Lu	**4f**
90 Th	91 Pa	92 U	93 Np	94 Pu	95 Am	96 Cm	97 Bk	98 Cf	99 Es	100 Fm	101 Md	102 No	103 Lr	**5f**

A periodic table colored to show the kinds of atomic orbitals (subshells) being filled in different parts of the periodic table. The atomic orbitals are given below the symbols of blocks of elements. The electronic structures of the A group elements are quite regular and can be predicted from their positions in the periodic table, but there are many exceptions in the *d* and *f* blocks. The populations of subshells are given in the table on pages A-6 and A-7.

Electron Configurations of the Atoms of the Elements

Element	Atomic Number	Populations of Subshells										
		1s	2s	2p	3s	3p	3d	4s	4p	4d	4f	5s
H	1	1										
He	2	2										
Li	3	2	1									
Be	4	2	2									
B	5	2	2	1								
C	6	2	2	2								
N	7	2	2	3								
O	8	2	2	4								
F	9	2	2	5								
Ne	10	2	2	6								
Na	11	Neon core			1							
Mg	12				2							
Al	13				2	1						
Si	14				2	2						
P	15				2	3						
S	16				2	4						
Cl	17				2	5						
Ar	18	2	2	6	2	6						
K	19	Argon core						1				
Ca	20							2				
Sc	21						1	2				
Ti	22						2	2				
V	23						3	2				
Cr	24						5	1				
Mn	25						5	2				
Fe	26						6	2				
Co	27						7	2				
Ni	28						8	2				
Cu	29						10	1				
Zn	30						10	2				
Ga	31						10	2	1			
Ge	32						10	2	2			
As	33						10	2	3			
Se	34						10	2	4			
Br	35						10	2	5			
Kr	36	2	2	6	2	6	10	2	6			
Rb	37	Krypton core										1
Sr	38											2
Y	39									1		2
Zr	40									2		2
Nb	41									4		1
Mo	42									5		1
Tc	43									5		2
Ru	44									7		1
Rh	45									8		1
Pd	46									10		
Ag	47									10		1
Cd	48									10		2

Element	Atomic Number		4d	4f	5s	5p	5d	5f	6s	6p	6d	7s
In	49		10		2	1						
Sn	50		10		2	2						
Sb	51		10		2	3						
Te	52		10		2	4						
I	53		10		2	5						
Xe	54		10		2	6						
Cs	55		10		2	6			1			
Ba	56		10		2	6			2			
La	57		10		2	6	1		2			
Ce	58		10	1	2	6	1		2			
Pr	59		10	3	2	6			2			
Nd	60		10	4	2	6			2			
Pm	61		10	5	2	6			2			
Sm	62		10	6	2	6			2			
Eu	63		10	7	2	6			2			
Gd	64		10	7	2	6	1		2			
Tb	65		10	9	2	6			2			
Dy	66		10	10	2	6			2			
Ho	67		10	11	2	6			2			
Er	68		10	12	2	6			2			
Tm	69		10	13	2	6			2			
Yb	70		10	14	2	6			2			
Lu	71		10	14	2	6	1		2			
Hf	72		10	14	2	6	2		2			
Ta	73		10	14	2	6	3		2			
W	74		10	14	2	6	4		2			
Re	75	Krypton core	10	14	2	6	5		2			
Os	76		10	14	2	6	6		2			
Ir	77		10	14	2	6	7		2			
Pt	78		10	14	2	6	9		1			
Au	79		10	14	2	6	10		1			
Hg	80		10	14	2	6	10		2			
Tl	81		10	14	2	6	10		2	1		
Pb	82		10	14	2	6	10		2	2		
Bi	83		10	14	2	6	10		2	3		
Po	84		10	14	2	6	10		2	4		
At	85		10	14	2	6	10		2	5		
Rn	86		10	14	2	6	10		2	6		
Fr	87		10	14	2	6	10		2	6		1
Ra	88		10	14	2	6	10		2	6		2
Ac	89		10	14	2	6	10		2	6	1	2
Th	90		10	14	2	6	10		2	6	2	2
Pa	91		10	14	2	6	10	2	2	6	1	2
U	92		10	14	2	6	10	3	2	6	1	2
Np	93		10	14	2	6	10	4	2	6	1	2
Pu	94		10	14	2	6	10	6	2	6		2
Am	95		10	14	2	6	10	7	2	6		2
Cm	96		10	14	2	6	10	7	2	6	1	2
Bk	97		10	14	2	6	10	9	2	6		2
Cf	98		10	14	2	6	10	10	2	6		2
Es	99		10	14	2	6	10	11	2	6		2
Fm	100		10	14	2	6	10	12	2	6		2
Md	101		10	14	2	6	10	13	2	6		2
No	102		10	14	2	6	10	14	2	6		2
Lr	103		10	14	2	6	10	14	2	6	1	2
Rf	104		10	14	2	6	10	14	2	6	2	2
Db	105		10	14	2	6	10	14	2	6	3	2
Sg	106		10	14	2	6	10	14	2	6	4	2
Bh	107		10	14	2	6	10	14	2	6	5	2
Hs	108		10	14	2	6	10	14	2	6	6	2
Mt	109		10	14	2	6	10	14	2	6	7	2

APPENDIX C

COMMON UNITS, EQUIVALENCES, AND CONVERSION FACTORS

C-1 FUNDAMENTAL UNITS OF THE SI SYSTEM

The metric system was implemented by the French National Assembly in 1790 and has been modified many times. The International System of Units, or *le Système International* (SI), represents an extension of the metric system. It was adopted by the eleventh General Conference of Weights and Measures in 1960 and has also been modified since. It is constructed from seven base units, each of which represents a particular physical quantity (Table I).

The first five units listed in Table I are particularly useful in general chemistry. They are defined as follows.

1. The *meter* is defined as the distance light travels in a vacuum in 1/299,792,458 second.

2. The *kilogram* represents the mass of a platinum–iridium block kept at the International Bureau of Weights and Measures at Sèvres, France.

3. The *second* was redefined in 1967 as the duration of 9,192,631,770 periods of a certain line in the microwave spectrum of cesium-133.

4. The *kelvin* is 1/273.16 of the temperature interval between absolute zero and the triple point of water.

5. The *mole* is the amount of substance that contains as many entities as there are atoms in exactly 0.012 kg of carbon-12 (12 g of ^{12}C atoms).

TABLE I	*SI Fundamental Units*	
Physical Quantity	**Name of Unit**	**Symbol**
length	meter	m
mass	kilogram	kg
time	second	s
temperature	kelvin	K
amount of substance	mole	mol
electric current	ampere	A
luminous intensity	candela	cd

Prefixes Used with Metric Units and SI Units

Decimal fractions and multiples of metric and SI units are designated by the prefixes listed in Table II. Those most commonly used in general chemistry are underlined.

TABLE II	**Traditional Metric and SI Prefixes**				
Factor	**Prefix**	**Symbol**	**Factor**	**Prefix**	**Symbol**
10^{12}	tera	T	10^{-1}	deci	d
10^9	giga	G	10^{-2}	centi	c
10^6	mega	M	10^{-3}	milli	m
10^3	kilo	k	10^{-6}	micro	μ
10^2	hecto	h	10^{-9}	nano	n
10^1	deka	da	10^{-12}	pico	p
			10^{-15}	femto	f
			10^{-18}	atto	a

C-2 DERIVED SI UNITS

In the International System of Units all physical quantities are represented by appropriate combinations of the base units listed in Table I. A list of the derived units frequently used in general chemistry is given in Table III.

TABLE III	**Derived SI Units**		
Physical Quantity	**Name of Unit**	**Symbol**	**Definition**
area	square meter	m^2	
volume	cubic meter	m^3	
density	kilogram per cubic meter	kg/m^3	
force	newton	N	$kg \cdot m/s^2$
pressure	pascal	Pa	N/m^2
energy	joule	J	$kg \cdot m^2/s^2$
electric charge	coulomb	C	$A \cdot s$
electric potential difference	volt	V	$J/(A \cdot s)$

Common Units of Mass and Weight

1 pound = 453.59 grams

1 pound = 453.59 grams = 0.45359 kilogram
1 kilogram = 1000 grams = 2.205 pounds
1 gram = 10 decigrams = 100 centigrams
     = 1000 milligrams
1 gram = 6.022×10^{23} atomic mass units
1 atomic mass unit = 1.6606×10^{-24} gram
1 short ton = 2000 pounds = 907.2 kilograms
1 long ton = 2240 pounds
1 metric tonne = 1000 kilograms = 2205 pounds

Common Units of Length

1 inch = 2.54 centimeters (exactly)

1 mile = 5280 feet = 1.609 kilometers

1 yard = 36 inches = 0.9144 meter

1 meter = 100 centimeters = 39.37 inches = 3.281 feet
    = 1.094 yards

1 kilometer = 1000 meters = 1094 yards = 0.6215 mile

1 Ångstrom = 1.0×10^{-8} centimeter = 0.10 nanometer
    = 1.0×10^{-10} meter = 3.937×10^{-9} inch

Common Units of Volume

1 quart = 0.9463 liter
1 liter = 1.056 quarts

1 liter = 1 cubic decimeter = 1000 cubic centimeters
    = 0.001 cubic meter

1 milliliter = 1 cubic centimeter = 0.001 liter
    = 1.056×10^{-3} quart

1 cubic foot = 28.316 liters = 29.902 quarts
    = 7.475 gallons

Common Units of Force* and Pressure

1 atmosphere = 760 millimeters of mercury
    = 1.01325×10^{5} pascals
    = 1.01325 bar
    = 14.70 pounds per square inch

1 bar = 10^{5} pascals = 0.98692 atm

1 torr = 1 millimeter of mercury

1 pascal = $1 \text{ kg/m} \cdot \text{s}^2 = 1 \text{ N/m}^2$

Force: 1 newton (N) = $1 \text{ kg} \cdot m/s^2$, i.e., the force that, when applied for 1 second, gives a 1-kilogram mass a velocity of 1 meter per second.

Common Units of Energy

1 joule = 1×10^{7} ergs

1 thermochemical calorie* = 4.184 joules = 4.184×10^{7} ergs
    = 4.129×10^{-2} liter-atmospheres
    = 2.612×10^{19} electron volts

1 erg = 1×10^{-7} joule = 2.3901×10^{-8} calorie

1 electron volt = 1.6022×10^{-19} joule = 1.6022×10^{-12} erg = 96.487 kJ mol[†]

1 liter-atmosphere = 24.217 calories = 101.325 joules = 1.01325×10^{9} ergs

1 British thermal unit = 1055.06 joules = 1.05506×10^{10} ergs = 252.2 calories

The amount of heat required to raise the temperature of one gram of water from 14.5°C to 15.5°C.

†*Note that the other units are per particle and must be multiplied by 6.022×10^{23} to be strictly comparable.*

PHYSICAL CONSTANTS

Quantity	Symbol	Traditional Units	SI Units
Acceleration of gravity	g	980.6 cm/s	9.806 m/s
Atomic mass unit ($\frac{1}{12}$ the mass of ^{12}C atom)	amu or u	1.6606×10^{-24} g	1.6606×10^{-27} kg
Avogadro's number	N	6.0221367×10^{23} particles/mol	6.0221367×10^{23} particles/mol
Bohr radius	a_0	0.52918 Å 5.2918×10^{-9} cm	5.2918×10^{-11} m
Boltzmann constant	k	1.3807×10^{-16} erg/K	1.3807×10^{-23} J/K
Charge-to-mass ratio of electron	e/m	1.75882×10^8 coulomb/g	1.75882×10^{11} C/kg
Electronic charge	e	1.60218×10^{-19} coulomb 4.8033×10^{-10} esu	1.60218×10^{-19} C
Electron rest mass	m_e	9.10940×10^{-28} g 0.00054858 amu	9.10940×10^{-31} kg
Faraday constant	F	96,485 coulombs/eq 23.06 kcal/volt·eq	96,485 C/mol e^- 96,485 J/V·mol e^-
Gas constant	R	$0.08206 \frac{L \cdot atm}{mol \cdot K}$ $1.987 \frac{cal}{mol \cdot K}$	$8.3145 \frac{kPa \cdot dm^3}{mol \cdot K}$ 8.3145 J/mol·K
Molar volume (STP)	V_m	22.414 L/mol	22.414×10^{-3} m³/mol 22.414 dm³/mol
Neutron rest mass	m_n	1.67495×10^{-24} g 1.008665 amu	1.67495×10^{-27} kg
Planck constant	h	6.6262×10^{-27} erg·s	6.6262×10^{-34} J·s
Proton rest mass	m_p	1.6726×10^{-24} g 1.007277 amu	1.6726×10^{-27} kg
Rydberg constant	R_∞	3.289×10^{15} cycles/s 2.1799×10^{-11} erg	1.0974×10^7 m⁻¹ 2.1799×10^{-18} J
Speed of light (in a vacuum)	c	$2.99792458 \times 10^{10}$ cm/s (186,281 miles/second)	2.99792458×10^8 m/s

$\pi = 3.1416$ 2.303 R = 4.576 cal/mol·K = 19.15 J/mol·K

$e = 2.71828$ 2.303 RT *(at 25°C)* = 1364 cal/mol = 5709 J/mol

ln X = 2.303 *log* X

APPENDIX E

SOME PHYSICAL CONSTANTS FOR A FEW COMMON SUBSTANCES

Specific Heats and Heat Capacities for Some Common Substances

Substance	Specific Heat (J/g·°C)	Molar Heat Capacity (J/mol·°C)
Al(s)	0.900	24.3
Ca(s)	0.653	26.2
Cu(s)	0.385	24.5
Fe(s)	0.444	24.8
Hg(ℓ)	0.138	27.7
H_2O(s), ice	2.09	37.7
H_2O(ℓ), water	4.18	75.3
H_2O(g), steam	2.03	36.4
C_6H_6(ℓ), benzene	1.74	136
C_6H_6(g), benzene	1.04	81.6
C_2H_5OH(ℓ), ethanol	2.46	113
C_2H_5OH(g), ethanol	0.954	420
$(C_2H_5)_2O$(ℓ), diethyl ether	3.74	172
$(C_2H_5)_2O$(g), diethyl ether	2.35	108

Heats of Transformation and Transformation Temperatures of Several Substances

Substance	mp (°C)	Heat of Fusion (J/g)	ΔH_{fus} (kJ/mol)	bp (°C)	Heat of Vaporization (J/g)	ΔH_{vap} (kJ/mol)
Al	658	395	10.6	2467	10520	284
Ca	851	233	9.33	1487	4030	162
Cu	1083	205	13.0	2595	4790	305
H_2O	0.0	334	6.02	100	2260	40.7
Fe	1530	267	14.9	2735	6340	354
Hg	−39	11	23.3	357	292	58.6
CH_4	−182	58.6	0.92	−164	—	—
C_2H_5OH	−117	109	5.02	78.3	855	39.3
C_6H_6	5.48	127	9.92	80.1	395	30.8
$(C_2H_5)_2O$	−116	97.9	7.66	35	351	26.0

Vapor Pressure of Water at Various Temperatures

Temperature (°C)	Vapor Pressure (torr)	Temperature (°C)	Vapor Pressure (torr)	Temperature (°C)	Vapor Pressure (torr)	Temperature (°C)	Vapor Pressure (torr)
−10	2.1	21	18.7	51	97.2	81	369.7
−9	2.3	22	19.8	52	102.1	82	384.9
−8	2.5	23	21.1	53	107.2	83	400.6
−7	2.7	24	22.4	54	112.5	84	416.8
−6	2.9	25	23.8	55	118.0	85	433.6
−5	3.2	26	25.2	56	123.8	86	450.9
−4	3.4	27	26.7	57	129.8	87	468.7
−3	3.7	28	28.3	58	136.1	88	487.1
−2	4.0	29	30.0	59	142.6	89	506.1
−1	4.3	30	31.8	60	149.4	90	525.8
0	4.6	31	33.7	61	156.4	91	546.1
1	4.9	32	35.7	62	163.8	92	567.0
2	5.3	33	37.7	63	171.4	93	588.6
3	5.7	34	39.9	64	179.3	94	610.9
4	6.1	35	42.2	65	187.5	95	633.9
5	6.5	36	44.6	66	196.1	96	657.6
6	7.0	37	47.1	67	205.0	97	682.1
7	7.5	38	49.7	68	214.2	98	707.3
8	8.0	39	52.4	69	223.7	99	733.2
9	8.6	40	55.3	70	233.7	100	760.0
10	9.2	41	58.3	71	243.9	101	787.6
11	9.8	42	61.5	72	254.6	102	815.9
12	10.5	43	64.8	73	265.7	103	845.1
13	11.2	44	68.3	74	277.2	104	875.1
14	12.0	45	71.9	75	289.1	105	906.1
15	12.8	46	75.7	76	301.4	106	937.9
16	13.6	47	79.6	77	314.1	107	970.6
17	14.5	48	83.7	78	327.3	108	1004.4
18	15.5	49	88.0	79	341.0	109	1038.9
19	16.5	50	92.5	80	355.1	110	1074.6
20	17.5						

APPENDIX F

IONIZATION CONSTANTS FOR WEAK ACIDS AT 25°C

Acid	Formula and Ionization Equation		K_a
Acetic	CH_3COOH	$\rightleftharpoons H^+ + CH_3COO^-$	1.8×10^{-5}
Arsenic	H_3AsO_4	$\rightleftharpoons H^+ + H_2AsO_4^-$	$2.5 \times 10^{-4} = K_{a1}$
	$H_2AsO_4^-$	$\rightleftharpoons H^+ + HAsO_4^{2-}$	$5.6 \times 10^{-8} = K_{a2}$
	$HAsO_4^{2-}$	$\rightleftharpoons H^+ + AsO_4^{3-}$	$3.0 \times 10^{-13} = K_{a3}$
Arsenous	H_3AsO_3	$\rightleftharpoons H^+ + H_2AsO_3^-$	$6.0 \times 10^{-10} = K_{a1}$
	$H_2AsO_3^-$	$\rightleftharpoons H^+ + HAsO_3^{2-}$	$3.0 \times 10^{-14} = K_{a2}$
Benzoic	C_6H_5COOH	$\rightleftharpoons H^+ + C_6H_5COO^-$	6.3×10^{-5}
Boric*	$B(OH)_3$	$\rightleftharpoons H^+ + BO(OH)_2^-$	$7.3 \times 10^{-10} = K_{a1}$
	$BO(OH)_2^-$	$\rightleftharpoons H^+ + BO_2(OH)^{2-}$	$1.8 \times 10^{-13} = K_{a2}$
	$BO_2(OH)^{2-}$	$\rightleftharpoons H^+ + BO_3^{3-}$	$1.6 \times 10^{-14} = K_{a3}$
Carbonic	H_2CO_3	$\rightleftharpoons H^+ + HCO_3^-$	$4.2 \times 10^{-7} = K_{a1}$
	HCO_3^-	$\rightleftharpoons H^+ + CO_3^{2-}$	$4.8 \times 10^{-11} = K_{a2}$
Citric	$C_3H_5O(COOH)_3$	$\rightleftharpoons H^+ + C_4H_5O_3(COOH)_2^-$	$7.4 \times 10^{-3} = K_{a1}$
	$C_4H_5O_3(COOH)_2^-$	$\rightleftharpoons H^+ + C_5H_5O_5COOH^{2-}$	$1.7 \times 10^{-5} = K_{a2}$
	$C_5H_5O_5COOH^{2-}$	$\rightleftharpoons H^+ + C_6H_5O_7^{3-}$	$7.4 \times 10^{-7} = K_{a3}$
Cyanic	$HOCN$	$\rightleftharpoons H^+ + OCN^-$	3.5×10^{-4}
Formic	$HCOOH$	$\rightleftharpoons H^+ + HCOO^-$	1.8×10^{-4}
Hydrazoic	HN_3	$\rightleftharpoons H^+ + N_3^-$	1.9×10^{-5}
Hydrocyanic	HCN	$\rightleftharpoons H^+ + CN^-$	4.0×10^{-10}
Hydrofluoric	HF	$\rightleftharpoons H^+ + F^-$	7.2×10^{-4}
Hydrogen peroxide	H_2O_2	$\rightleftharpoons H^+ + HO_2^-$	2.4×10^{-12}
Hydrosulfuric	H_2S	$\rightleftharpoons H^+ + HS^-$	$1.0 \times 10^{-7} = K_{a1}$
	HS^-	$\rightleftharpoons H^+ + S^{2-}$	$1.0 \times 10^{-19} = K_{a2}$
Hypobromous	$HOBr$	$\rightleftharpoons H^+ + OBr^-$	2.5×10^{-9}
Hypochlorous	$HOCl$	$\rightleftharpoons H^+ + OCl^-$	3.5×10^{-8}
Hypoiodous	HOI	$\rightleftharpoons H^+ + OI^-$	2.3×10^{-11}
Nitrous	HNO_2	$\rightleftharpoons H^+ + NO_2^-$	4.5×10^{-4}
Oxalic	$(COOH)_2$	$\rightleftharpoons H^+ + COOCOOH^-$	$5.9 \times 10^{-2} = K_{a1}$
	$COOCOOH^-$	$\rightleftharpoons H^+ + (COO)_2^{2-}$	$6.4 \times 10^{-5} = K_{a2}$
Phenol	HC_6H_5O	$\rightleftharpoons H^+ + C_6H_5O^-$	1.3×10^{-10}
Phosphoric	H_3PO_4	$\rightleftharpoons H^+ + H_2PO_4^-$	$7.5 \times 10^{-3} = K_{a1}$
	$H_2PO_4^-$	$\rightleftharpoons H^+ + HPO_4^{2-}$	$6.2 \times 10^{-8} = K_{a2}$
	HPO_4^{2-}	$\rightleftharpoons H^+ + PO_4^{3-}$	$3.6 \times 10^{-13} = K_{a3}$
Phosphorous	H_3PO_3	$\rightleftharpoons H^+ + H_2PO_3^-$	$1.6 \times 10^{-2} = K_{a1}$
	$H_2PO_3^-$	$\rightleftharpoons H^+ + HPO_3^{2-}$	$7.0 \times 10^{-7} = K_{a2}$
Selenic	H_2SeO_4	$\rightleftharpoons H^+ + HSeO_4^-$	Very large $= K_{a1}$
	$HSeO_4^-$	$\rightleftharpoons H^+ + SeO_4^{2-}$	$1.2 \times 10^{-2} = K_{a2}$
Selenous	H_2SeO_3	$\rightleftharpoons H^+ + HSeO_3^-$	$2.7 \times 10^{-3} = K_{a1}$
	$HSeO_3^-$	$\rightleftharpoons H^+ + SeO_3^{2-}$	$2.5 \times 10^{-7} = K_{a2}$
Sulfuric	H_2SO_4	$\rightleftharpoons H^+ + HSO_4^-$	Very large $= K_{a1}$
	HSO_4^-	$\rightleftharpoons H^+ + SO_4^{2-}$	$1.2 \times 10^{-2} = K_{a2}$
Sulfurous	H_2SO_3	$\rightleftharpoons H^+ + HSO_3^-$	$1.2 \times 10^{-2} = K_{a1}$
	HSO_3^-	$\rightleftharpoons H^+ + SO_3^{2-}$	$6.2 \times 10^{-8} = K_{a2}$
Tellurous	H_2TeO_3	$\rightleftharpoons H^+ + HTeO_3^-$	$2 \times 10^{-3} = K_{a1}$
	$HTeO_3^-$	$\rightleftharpoons H^+ + TeO_3^{2-}$	$1 \times 10^{-8} = K_{a2}$

Boric acid acts as a Lewis acid in aqueous solution.

IONIZATION CONSTANTS FOR WEAK BASES AT 25°C

Base	Formula and Ionization Equation				K_b
Ammonia	NH_3	$+ H_2O \rightleftharpoons NH_4^+$		$+ OH^-$	1.8×10^{-5}
Aniline	$C_6H_5NH_2$	$+ H_2O \rightleftharpoons C_6H_5NH_3^+$		$+ OH^-$	4.2×10^{-10}
Dimethylamine	$(CH_3)_2NH$	$+ H_2O \rightleftharpoons (CH_3)_2NH_2^+$		$+ OH^-$	7.4×10^{-4}
Ethylenediamine	$(CH_2)_2(NH_2)_2$	$+ H_2O \rightleftharpoons (CH_2)_2(NH_2)_2H^+$		$+ OH^-$	$8.5 \times 10^{-5} = K_{b1}$
	$(CH_2)_2(NH_2)_2H^+$	$+ H_2O \rightleftharpoons (CH_2)_2(NH_2)_2H_2^{2+}$		$+ OH^-$	$2.7 \times 10^{-8} = K_{b2}$
Hydrazine	N_2H_4	$+ H_2O \rightleftharpoons N_2H_5^+$		$+ OH^-$	$8.5 \times 10^{-7} = K_{b1}$
	$N_2H_5^+$	$+ H_2O \rightleftharpoons N_2H_6^{2+}$		$+ OH^-$	$8.9 \times 10^{-16} = K_{b2}$
Hydroxylamine	NH_2OH	$+ H_2O \rightleftharpoons NH_3OH^+$		$+ OH^-$	6.6×10^{-9}
Methylamine	CH_3NH_2	$+ H_2O \rightleftharpoons CH_3NH_3^+$		$+ OH^-$	5.0×10^{-4}
Pyridine	C_5H_5N	$+ H_2O \rightleftharpoons C_5H_5NH^+$		$+ OH^-$	1.5×10^{-9}
Trimethylamine	$(CH_3)_3N$	$+ H_2O \rightleftharpoons (CH_3)_3NH^+$		$+ OH^-$	7.4×10^{-5}

APPENDIX H

SOLUBILITY PRODUCT CONSTANTS FOR SOME INORGANIC COMPOUNDS AT 25°C

Substance	K_{sp}	Substance	K_{sp}
Aluminum compounds		*Chromium compounds*	
$AlAsO_4$	1.6×10^{-16}	$CrAsO_4$	7.8×10^{-21}
$Al(OH)_3$	1.9×10^{-33}	$Cr(OH)_3$	6.7×10^{-31}
$AlPO_4$	1.3×10^{-20}	$CrPO_4$	2.4×10^{-23}
Antimony compounds		*Cobalt compounds*	
Sb_2S_3	1.6×10^{-93}	$Co_3(AsO_4)_2$	7.6×10^{-29}
Barium compounds		$CoCO_3$	8.0×10^{-13}
$Ba_3(AsO_4)_2$	1.1×10^{-13}	$Co(OH)_2$	2.5×10^{-16}
$BaCO_3$	8.1×10^{-9}	$CoS\ (\alpha)$	5.9×10^{-21}
$BaC_2O_4 \cdot 2H_2O^*$	1.1×10^{-7}	$CoS\ (\beta)$	8.7×10^{-23}
$BaCrO_4$	2.0×10^{-10}	$Co(OH)_3$	4.0×10^{-45}
BaF_2	1.7×10^{-6}	Co_2S_3	2.6×10^{-124}
$Ba(OH)_2 \cdot 8H_2O^*$	5.0×10^{-3}	*Copper compounds*	
$Ba_3(PO_4)_2$	1.3×10^{-29}	$CuBr$	5.3×10^{-9}
$BaSeO_4$	2.8×10^{-11}	$CuCl$	1.9×10^{-7}
$BaSO_3$	8.0×10^{-7}	$CuCN$	3.2×10^{-20}
$BaSO_4$	1.1×10^{-10}	$Cu_2O\ (Cu^+ + OH^-)^\dagger$	1.0×10^{-14}
Bismuth compounds		CuI	5.1×10^{-12}
$BiOCl$	7.0×10^{-9}	Cu_2S	1.6×10^{-48}
$BiO(OH)$	1.0×10^{-12}	$CuSCN$	1.6×10^{-11}
$Bi(OH)_3$	3.2×10^{-40}	$Cu_3(AsO_4)_2$	7.6×10^{-36}
BiI_3	8.1×10^{-19}	$CuCO_3$	2.5×10^{-10}
$BiPO_4$	1.3×10^{-23}	$Cu_2[Fe(CN)_6]$	1.3×10^{-16}
Bi_2S_3	1.6×10^{-72}	$Cu(OH)_2$	1.6×10^{-19}
Cadmium compounds		CuS	8.7×10^{-36}
$Cd_3(AsO_4)_2$	2.2×10^{-32}	*Gold compounds*	
$CdCO_3$	2.5×10^{-14}	$AuBr$	5.0×10^{-17}
$Cd(CN)_2$	1.0×10^{-8}	$AuCl$	2.0×10^{-13}
$Cd_2[Fe(CN)_6]$	3.2×10^{-17}	AuI	1.6×10^{-23}
$Cd(OH)_2$	1.2×10^{-14}	$AuBr_3$	4.0×10^{-36}
CdS	3.6×10^{-29}	$AuCl_3$	3.2×10^{-25}
Calcium compounds		$Au(OH)_3$	1.0×10^{-53}
$Ca_3(AsO_4)_2$	6.8×10^{-19}	AuI_3	1.0×10^{-46}
$CaCO_3$	4.8×10^{-9}	*Iron compounds*	
$CaCrO_4$	7.1×10^{-4}	$FeCO_3$	3.5×10^{-11}
$CaC_2O_4 \cdot H_2O^*$	2.3×10^{-9}	$Fe(OH)_2$	7.9×10^{-15}
CaF_2	3.9×10^{-11}	FeS	4.9×10^{-18}
$Ca(OH)_2$	7.9×10^{-6}	$Fe_4[Fe(CN)_6]_3$	3.0×10^{-41}
$CaHPO_4$	2.7×10^{-7}	$Fe(OH)_3$	6.3×10^{-38}
$Ca(H_2PO_4)_2$	1.0×10^{-3}	Fe_2S_3	1.4×10^{-88}
$Ca_3(PO_4)_2$	1.0×10^{-25}	*Lead compounds*	
$CaSO_3 \cdot 2H_2O^*$	1.3×10^{-8}	$Pb_3(AsO_4)_2$	4.1×10^{-36}
$CaSO_4 \cdot 2H_2O^*$	2.4×10^{-5}	$PbBr_2$	6.3×10^{-6}

SOLUBILITY PRODUCT CONSTANTS FOR SOME INORGANIC COMPOUNDS AT 25°C (continued)

Substance	K_{sp}	Substance	K_{sp}
Lead compounds (cont.)		Nickel compounds (cont.)	
$PbCO_3$	1.5×10^{-13}	$NiS (\alpha)$	3.0×10^{-21}
$PbCl_2$	1.7×10^{-5}	$NiS (\beta)$	1.0×10^{-26}
$PbCrO_4$	1.8×10^{-14}	$NiS (\gamma)$	2.0×10^{-28}
PbF_2	3.7×10^{-8}	Silver compounds	
$Pb(OH)_2$	2.8×10^{-16}	Ag_3AsO_4	1.1×10^{-20}
PbI_2	8.7×10^{-9}	$AgBr$	3.3×10^{-13}
$Pb_3(PO_4)_2$	3.0×10^{-44}	Ag_2CO_3	8.1×10^{-12}
$PbSeO_4$	1.5×10^{-7}	$AgCl$	1.8×10^{-10}
$PbSO_4$	1.8×10^{-8}	Ag_2CrO_4	9.0×10^{-12}
PbS	8.4×10^{-28}	$AgCN$	1.2×10^{-16}
Magnesium compounds		$Ag_4[Fe(CN)_6]$	1.6×10^{-41}
$Mg_3(AsO_4)_2$	2.1×10^{-20}	$Ag_2O (Ag^+ + OH^-)^\dagger$	2.0×10^{-8}
$MgCO_3 \cdot 3H_2O^*$	4.0×10^{-5}	AgI	1.5×10^{-16}
MgC_2O_4	8.6×10^{-5}	Ag_3PO_4	1.3×10^{-20}
MgF_2	6.4×10^{-9}	Ag_2SO_3	1.5×10^{-14}
$Mg(OH)_2$	1.5×10^{-11}	Ag_2SO_4	1.7×10^{-5}
$MgNH_4PO_4$	2.5×10^{-12}	Ag_2S	1.0×10^{-49}
Manganese compounds		$AgSCN$	1.0×10^{-12}
$Mn_3(AsO_4)_2$	1.9×10^{-11}	Strontium compounds	
$MnCO_3$	1.8×10^{-11}	$Sr_3(AsO_4)_2$	1.3×10^{-18}
$Mn(OH)_2$	4.6×10^{-14}	$SrCO_3$	9.4×10^{-10}
MnS	5.1×10^{-15}	$SrC_2O_4 \cdot 2H_2O^*$	5.6×10^{-8}
$Mn(OH)_3$	$\approx 1.0 \times 10^{-36}$	$SrCrO_4$	3.6×10^{-5}
Mercury compounds		$Sr(OH)_2 \cdot 8H_2O^*$	3.2×10^{-4}
Hg_2Br_2	1.3×10^{-22}	$Sr_3(PO_4)_2$	1.0×10^{-31}
Hg_2CO_3	8.9×10^{-17}	$SrSO_3$	4.0×10^{-8}
Hg_2Cl_2	1.1×10^{-18}	$SrSO_4$	2.8×10^{-7}
Hg_2CrO_4	5.0×10^{-9}	Tin compounds	
Hg_2I_2	4.5×10^{-29}	$Sn(OH)_2$	2.0×10^{-26}
$Hg_2O \cdot H_2O^*$		SnI_2	1.0×10^{-4}
$(Hg_2^{2+} + 2OH^-)^\dagger$	1.6×10^{-23}	SnS	1.0×10^{-28}
Hg_2SO_4	6.8×10^{-7}	$Sn(OH)_4$	1.0×10^{-57}
Hg_2S	5.8×10^{-44}	SnS_2	1.0×10^{-70}
$Hg(CN)_2$	3.0×10^{-23}	Zinc compounds	
$Hg(OH)_2$	2.5×10^{-26}	$Zn_3(AsO_4)_2$	1.1×10^{-27}
HgI_2	4.0×10^{-29}	$ZnCO_3$	1.5×10^{-11}
HgS	3.0×10^{-53}	$Zn(CN)_2$	8.0×10^{-12}
Nickel compounds		$Zn_2[Fe(CN)_6]$	4.1×10^{-16}
$Ni_3(AsO_4)_2$	1.9×10^{-26}	$Zn(OH)_2$	4.5×10^{-17}
$NiCO_3$	6.6×10^{-9}	$Zn_3(PO_4)_2$	9.1×10^{-33}
$Ni(CN)_2$	3.0×10^{-23}	ZnS	1.1×10^{-21}
$Ni(OH)_2$	2.8×10^{-16}		

*$[H_2O]$ does not appear in equilibrium constants for equilibria in aqueous solution in general, so it does not appear in the K_{sp} expressions for hydrated solids.

†Very small amounts of oxides dissolve in water to give the ions indicated in parentheses. These solid hydroxides are unstable and decompose to oxides as rapidly as they are formed.

APPENDIX I

DISSOCIATION CONSTANTS FOR SOME COMPLEX IONS

Dissociation Equilibrium	K_d
$[AgBr_2]^- \rightleftharpoons Ag^+ + 2Br^-$	7.8×10^{-8}
$[AgCl_2]^- \rightleftharpoons Ag^+ + 2Cl^-$	4.0×10^{-6}
$[Ag(CN)_2]^- \rightleftharpoons Ag^+ + 2CN^-$	1.8×10^{-19}
$[Ag(S_2O_3)_2]^{3-} \rightleftharpoons Ag^+ + 2S_2O_3^{2-}$	5.0×10^{-14}
$[Ag(NH_3)_2]^+ \rightleftharpoons Ag^+ + 2NH_3$	6.3×10^{-8}
$[Ag(en)]^+ \rightleftharpoons Ag^+ + en^*$	1.0×10^{-5}
$[AlF_6]^{3-} \rightleftharpoons Al^{3+} + 6F^-$	2.0×10^{-24}
$[Al(OH)_4]^- \rightleftharpoons Al^{3+} + 4OH^-$	1.3×10^{-34}
$[Au(CN)_2]^- \rightleftharpoons Au^+ + 2CN^-$	5.0×10^{-39}
$[Cd(CN)_4]^{2-} \rightleftharpoons Cd^{2+} + 4CN^-$	7.8×10^{-18}
$[CdCl_4]^{2-} \rightleftharpoons Cd^{2+} + 4Cl^-$	1.0×10^{-4}
$[Cd(NH_3)_4]^{2+} \rightleftharpoons Cd^{2+} + 4NH_3$	1.0×10^{-7}
$[Co(NH_3)_6]^{2+} \rightleftharpoons Co^{2+} + 6NH_3$	1.3×10^{-5}
$[Co(NH_3)_6]^{3+} \rightleftharpoons Co^{3+} + 6NH_3$	2.2×10^{-34}
$[Co(en)_3]^{2+} \rightleftharpoons Co^{2+} + 3en^*$	1.5×10^{-14}
$[Co(en)_3]^{3+} \rightleftharpoons Co^{3+} + 3en^*$	2.0×10^{-49}
$[Cu(CN)_2]^- \rightleftharpoons Cu^+ + 2CN^-$	1.0×10^{-16}
$[CuCl_2]^- \rightleftharpoons Cu^+ + 2Cl^-$	1.0×10^{-5}
$[Cu(NH_3)_2]^+ \rightleftharpoons Cu^+ + 2NH_3$	1.4×10^{-11}
$[Cu(NH_3)_4]^{2+} \rightleftharpoons Cu^{2+} + 4NH_3$	8.5×10^{-13}
$[Fe(CN)_6]^{4-} \rightleftharpoons Fe^{2+} + 6CN^-$	1.3×10^{-37}
$[Fe(CN)_6]^{3-} \rightleftharpoons Fe^{3+} + 6CN^-$	1.3×10^{-44}
$[HgCl_4]^{2-} \rightleftharpoons Hg^{2+} + 4Cl^-$	8.3×10^{-16}
$[Ni(CN)_4]^{2-} \rightleftharpoons Ni^{2+} + 4CN^-$	1.0×10^{-31}
$[Ni(NH_3)_6]^{2+} \rightleftharpoons Ni^{2+} + 6NH_3$	1.8×10^{-9}
$[Zn(OH)_4]^{2-} \rightleftharpoons Zn^{2+} + 4OH^-$	3.5×10^{-16}
$[Zn(NH_3)_4]^{2+} \rightleftharpoons Zn^{2+} + 4NH_3$	3.4×10^{-10}

*The abbreviation "en" represents ethylenediamine, $H_2NCH_2CH_2NH_2$.

STANDARD REDUCTION POTENTIALS IN AQUEOUS SOLUTION AT 25°C

Acidic Solution	Standard Reduction Potential, E^0 (volts)
$Li^+(aq) + e^- \longrightarrow Li(s)$	-3.045
$K^+(aq) + e^- \longrightarrow K(s)$	-2.925
$Rb^+(aq) + e^- \longrightarrow Rb(s)$	-2.925
$Ba^{2+}(aq) + 2e^- \longrightarrow Ba(s)$	-2.90
$Sr^{2+}(aq) + 2e^- \longrightarrow Sr(s)$	-2.89
$Ca^{2+}(aq) + 2e^- \longrightarrow Ca(s)$	-2.87
$Na^+(aq) + e^- \longrightarrow Na(s)$	-2.714
$Mg^{2+}(aq) + 2e^- \longrightarrow Mg(s)$	-2.37
$H_2(g) + 2e^- \longrightarrow 2H^-(aq)$	-2.25
$Al^{3+}(aq) + 3e^- \longrightarrow Al(s)$	-1.66
$Zr^{4+}(aq) + 4e^- \longrightarrow Zr(s)$	-1.53
$ZnS(s) + 2e^- \longrightarrow Zn(s) + S^{2-}(aq)$	-1.44
$CdS(s) + 2e^- \longrightarrow Cd(s) + S^{2-}(aq)$	-1.21
$V^{2+}(aq) + 2e^- \longrightarrow V(s)$	-1.18
$Mn^{2+}(aq) + 2e^- \longrightarrow Mn(s)$	-1.18
$FeS(s) + 2e^- \longrightarrow Fe(s) + S^{2-}(aq)$	-1.01
$Cr^{2+}(aq) + 2e^- \longrightarrow Cr(s)$	-0.91
$Zn^{2+}(aq) + 2e^- \longrightarrow Zn(s)$	-0.763
$Cr^{3+}(aq) + 3e^- \longrightarrow Cr(s)$	-0.74
$HgS(s) + 2H^+(aq) + 2e^- \longrightarrow Hg(\ell) + H_2S(g)$	-0.72
$Ga^{3+}(aq) + 3e^- \longrightarrow Ga(s)$	-0.53
$2CO_2(g) + 2H^+(aq) + 2e^- \longrightarrow (COOH)_2(aq)$	-0.49
$Fe^{2+}(aq) + 2e^- \longrightarrow Fe(s)$	-0.44
$Cr^{3+}(aq) + e^- \longrightarrow Cr^{2+}(aq)$	-0.41
$Cd^{2+}(aq) + 2e^- \longrightarrow Cd(s)$	-0.403
$Se(s) + 2H^+(aq) + 2e^- \longrightarrow H_2Se(aq)$	-0.40
$PbSO_4(s) + 2e^- \longrightarrow Pb(s) + SO_4^{2-}(aq)$	-0.356
$Tl^+(aq) + e^- \longrightarrow Tl(s)$	-0.34
$Co^{2+}(aq) + 2e^- \longrightarrow Co(s)$	-0.28
$Ni^{2+}(aq) + 2e^- \longrightarrow Ni(s)$	-0.25
$[SnF_6]^{2-}(aq) + 4e^- \longrightarrow Sn(s) + 6F^-(aq)$	-0.25
$AgI(s) + e^- \longrightarrow Ag(s) + I^-(aq)$	-0.15
$Sn^{2+}(aq) + 2e^- \longrightarrow Sn(s)$	-0.14
$Pb^{2+}(aq) + 2e^- \longrightarrow Pb(s)$	-0.126
$N_2O(g) + 6H^+(aq) + H_2O + 4e^- \longrightarrow 2NH_3OH^+(aq)$	-0.05
$2H^+(aq) + 2e^- \longrightarrow H_2(g)$ (reference electrode)	0.000
$AgBr(s) + e^- \longrightarrow Ag(s) + Br^-(aq)$	0.10
$S(s) + 2H^+(aq) + 2e^- \longrightarrow H_2S(aq)$	0.14
$Sn^{4+}(aq) + 2e^- \longrightarrow Sn^{2+}(aq)$	0.15
$Cu^{2+}(aq) + e^- \longrightarrow Cu^+(aq)$	0.153
$SO_4^{2-}(aq) + 4H^+(aq) + 2e^- \longrightarrow H_2SO_3(aq) + H_2O$	0.17

STANDARD REDUCTION POTENTIALS IN AQUEOUS SOLUTION AT 25°C (continued)

Acidic Solution	Standard Reduction Potential, E^0 (volts)
$SO_4^{2-}(aq) + 4H^+(aq) + 2e^- \longrightarrow SO_2(g) + 2H_2O$	0.20
$AgCl(s) + e^- \longrightarrow Ag(s) + Cl^-(aq)$	0.222
$Hg_2Cl_2(s) + 2e^- \longrightarrow 2Hg(\ell) + 2Cl^-(aq)$	0.27
$Cu^{2+}(aq) + 2e^- \longrightarrow Cu(s)$	0.337
$[RhCl_6]^{3-}(aq) + 3e^- \longrightarrow Rh(s) + 6Cl^-(aq)$	0.44
$Cu^+(aq) + e^- \longrightarrow Cu(s)$	0.521
$TeO_2(s) + 4H^+(aq) + 4e^- \longrightarrow Te(s) + 2H_2O$	0.529
$I_2(s) + 2e^- \longrightarrow 2I^-(aq)$	0.535
$H_3AsO_4(aq) + 2H^+(aq) + 2e^- \longrightarrow H_3AsO_3(aq) + H_2O$	0.58
$[PtCl_6]^{2-}(aq) + 2e^- \longrightarrow [PtCl_4]^{2-}(aq) + 2Cl^-(aq)$	0.68
$O_2(g) + 2H^+(aq) + 2e^- \longrightarrow H_2O_2(aq)$	0.682
$[PtCl_4]^{2-}(aq) + 2e^- \longrightarrow Pt(s) + 4Cl^-(aq)$	0.73
$SbCl_6^-(aq) + 2e^- \longrightarrow SbCl_4^-(aq) + 2Cl^-(aq)$	0.75
$Fe^{3+}(aq) + e^- \longrightarrow Fe^{2+}(aq)$	0.771
$Hg_2^{2+}(aq) + 2e^- \longrightarrow 2Hg(\ell)$	0.789
$Ag^+(aq) + e^- \longrightarrow Ag(s)$	0.7994
$Hg^{2+}(aq) + 2e^- \longrightarrow Hg(\ell)$	0.855
$2Hg^{2+}(aq) + 2e^- \longrightarrow Hg_2^{2+}(aq)$	0.920
$NO_3^-(aq) + 3H^+(aq) + 2e^- \longrightarrow HNO_2(aq) + H_2O$	0.94
$NO_3^-(aq) + 4H^+(aq) + 3e^- \longrightarrow NO(g) + 2H_2O$	0.96
$Pd^{2+}(aq) + 2e^- \longrightarrow Pd(s)$	0.987
$AuCl_4^-(aq) + 3e^- \longrightarrow Au(s) + 4Cl^-(aq)$	1.00
$Br_2(\ell) + 2e^- \longrightarrow 2Br^-(aq)$	1.08
$ClO_4^-(aq) + 2H^+(aq) + 2e^- \longrightarrow ClO_3^-(aq) + H_2O$	1.19
$IO_3^-(aq) + 6H^+(aq) + 5e^- \longrightarrow \frac{1}{2}I_2(aq) + 3H_2O$	1.195
$Pt^{2+}(aq) + 2e^- \longrightarrow Pt(s)$	1.2
$O_2(g) + 4H^+(aq) + 4e^- \longrightarrow 2H_2O$	1.229
$MnO_2(s) + 4H^+(aq) + 2e^- \longrightarrow Mn^{2+}(aq) + 2H_2O$	1.23
$N_2H_5^+(aq) + 3H^+(aq) + 2e^- \longrightarrow 2NH_4^+(aq)$	1.24
$Cr_2O_7^{2-}(aq) + 14H^+(aq) + 6e^- \longrightarrow 2Cr^{3+}(aq) + 7H_2O$	1.33
$Cl_2(g) + 2e^- \longrightarrow 2Cl^-(aq)$	1.360
$BrO_3^-(aq) + 6H^+(aq) + 6e^- \longrightarrow Br^-(aq) + 3H_2O$	1.44
$ClO_3^-(aq) + 6H^+(aq) + 5e^- \longrightarrow \frac{1}{2}Cl_2(g) + 3H_2O$	1.47
$Au^{3+}(aq) + 3e^- \longrightarrow Au(s)$	1.50
$MnO_4^-(aq) + 8H^+(aq) + 5e^- \longrightarrow Mn^{2+}(aq) + 4H_2O$	1.507
$NaBiO_3(s) + 6H^+(aq) + 2e^- \longrightarrow Bi^{3+}(aq) + Na^+(aq) + 3H_2O$	1.6
$Ce^{4+}(aq) + e^- \longrightarrow Ce^{3+}(aq)$	1.61
$2HOCl(aq) + 2H^+(aq) + 2e^- \longrightarrow Cl_2(g) + 2H_2O$	1.63
$Au^+(aq) + e^- \longrightarrow Au(s)$	1.68
$PbO_2(s) + SO_4^{2-}(aq) + 4H^+(aq) + 2e^- \longrightarrow PbSO_4(s) + 2H_2O$	1.685
$NiO_2(s) + 4H^+(aq) + 2e^- \longrightarrow Ni^{2+}(aq) + 2H_2O$	1.7
$H_2O_2(aq) + 2H^+(aq) + 2e^- \longrightarrow 2H_2O$	1.77
$Pb^{4+}(aq) + 2e^- \longrightarrow Pb^{2+}(aq)$	1.8
$Co^{3+}(aq) + e^- \longrightarrow Co^{2+}(aq)$	1.82
$F_2(g) + 2e^- \longrightarrow 2F^-(aq)$	2.87

STANDARD REDUCTION POTENTIALS IN AQUEOUS SOLUTION AT 25°C *(continued)*

Basic Solution	Standard Reduction Potential, E^0 (volts)
$SiO_3^{2-}(aq) + 3H_2O + 4e^- \longrightarrow Si(s) + 6OH^-(aq)$	-1.70
$Cr(OH)_3(s) + 3e^- \longrightarrow Cr(s) + 3OH^-(aq)$	-1.30
$[Zn(CN)_4]^{2-}(aq) + 2e^- \longrightarrow Zn(s) + 4CN^-(aq)$	-1.26
$Zn(OH)_2(s) + 2e^- \longrightarrow Zn(s) + 2OH^-(aq)$	-1.245
$[Zn(OH)_4]^{2-}(aq) + 2e^- \longrightarrow Zn(s) + 4OH^-(aq)$	-1.22
$N_2(g) + 4H_2O + 4e^- \longrightarrow N_2H_4(aq) + 4OH^-(aq)$	-1.15
$SO_4^{2-}(aq) + H_2O + 2e^- \longrightarrow SO_3^{2-}(aq) + 2OH^-(aq)$	-0.93
$Fe(OH)_2(s) + 2e^- \longrightarrow Fe(s) + 2OH^-(aq)$	-0.877
$2NO_3^-(aq) + 2H_2O + 2e^- \longrightarrow N_2O_4(g) + 4OH^-(aq)$	-0.85
$2H_2O + 2e^- \longrightarrow H_2(g) + 2OH^-(aq)$	-0.828
$Fe(OH)_3(s) + e^- \longrightarrow Fe(OH)_2(s) + OH^-(aq)$	-0.56
$S(s) + 2e^- \longrightarrow S^{2-}(aq)$	-0.48
$Cu(OH)_2(s) + 2e^- \longrightarrow Cu(s) + 2OH^-(aq)$	-0.36
$CrO_4^{2-}(aq) + 4H_2O + 3e^- \longrightarrow Cr(OH)_3(s) + 5OH^-(aq)$	-0.12
$MnO_2(s) + 2H_2O + 2e^- \longrightarrow Mn(OH)_2(s) + 2OH^-(aq)$	-0.05
$NO_3^-(aq) + H_2O + 2e^- \longrightarrow NO_2^-(aq) + 2OH^-(aq)$	0.01
$O_2(g) + H_2O + 2e^- \longrightarrow OOH^-(aq) + OH^-(aq)$	0.076
$HgO(s) + H_2O + 2e^- \longrightarrow Hg(\ell) + 2OH^-(aq)$	0.0984
$[Co(NH_3)_6]^{3+}(aq) + e^- \longrightarrow [Co(NH_3)_6]^{2+}(aq)$	0.10
$N_2H_4(aq) + 2H_2O + 2e^- \longrightarrow 2NH_3(aq) + 2OH^-(aq)$	0.10
$2NO_2^-(aq) + 3H_2O + 4e^- \longrightarrow N_2O(g) + 6OH^-(aq)$	0.15
$Ag_2O(s) + H_2O + 2e^- \longrightarrow 2Ag(s) + 2OH^-(aq)$	0.34
$ClO_4^-(aq) + H_2O + 2e^- \longrightarrow ClO_3^-(aq) + 2OH^-(aq)$	0.36
$O_2(g) + 2H_2O + 4e^- \longrightarrow 4OH^-(aq)$	0.40
$Ag_2CrO_4(s) + 2e^- \longrightarrow 2Ag(s) + CrO_4^{2-}(aq)$	0.446
$NiO_2(s) + 2H_2O + 2e^- \longrightarrow Ni(OH)_2(s) + 2OH^-(aq)$	0.49
$MnO_4^-(aq) + e^- \longrightarrow MnO_4^{2-}(aq)$	0.564
$MnO_4^-(aq) + 2H_2O + 3e^- \longrightarrow MnO_2(s) + 4OH^-(aq)$	0.588
$ClO_3^-(aq) + 3H_2O + 6e^- \longrightarrow Cl^-(aq) + 6OH^-(aq)$	0.62
$2NH_2OH(aq) + 2e^- \longrightarrow N_2H_4(aq) + 2OH^-(aq)$	0.74
$OOH^-(aq) + H_2O + 2e^- \longrightarrow 3OH^-(aq)$	0.88
$ClO^-(aq) + H_2O + 2e^- \longrightarrow Cl^-(aq) + 2OH^-(aq)$	0.89

APPENDIX K

SELECTED THERMODYNAMIC VALUES AT 298.15 K

Species	ΔH_f^0 (kJ/mol)	S^0 (J/mol·K)	ΔG_f^0 (kJ/mol)	Species	ΔH_f^0 (kJ/mol)	S^0 (J/mol·K)	ΔG_f^0 (kJ/mol)
Aluminum				*Cesium*			
Al(s)	0	28.3	0	Cs^+(aq)	−248	133	−282.0
$AlCl_3$(s)	−704.2	110.7	−628.9	CsF(aq)	−568.6	123	−558.5
Al_2O_3(s)	−1676	50.92	−1582	*Chlorine*			
Barium				Cl(g)	121.7	165.1	105.7
$BaCl_2$(s)	−860.1	126	−810.9	Cl^-(g)	−226	—	—
$BaSO_4$(s)	−1465	132	−1353	Cl_2(g)	0	223.0	0
Beryllium				HCl(g)	−92.31	186.8	−95.30
Be(s)	0	9.54	0	HCl(aq)	−167.4	55.10	−131.2
$Be(OH)_2$(s)	−907.1	—	—	*Chromium*			
Bromine				Cr(s)	0	23.8	0
Br(g)	111.8	174.9	82.4	$(NH_4)_2Cr_2O_7$(s)	−1807	—	—
$Br_2(\ell)$	0	152.23	0	*Copper*			
Br_2(g)	30.91	245.4	3.14	Cu(s)	0	33.15	0
BrF_3(g)	−255.6	292.4	−229.5	CuO(s)	−157	42.63	−130
HBr(g)	−36.4	198.59	−53.43	*Fluorine*			
Calcium				F^-(g)	−322	—	—
Ca(s)	0	41.6	0	F^-(aq)	−332.6	—	−278.8
Ca(g)	192.6	154.8	158.9	F(g)	78.99	158.6	61.92
Ca^{2+}(g)	1920	—	—	F_2(g)	0	202.7	0
CaC_2(s)	−62.8	70.3	−67.8	HF(g)	−271	173.7	−273
$CaCO_3$(s)	−1207	92.9	−1129	HF(aq)	−320.8	—	−296.8
$CaCl_2$(s)	−795.0	114	−750.2	*Hydrogen*			
CaF_2(s)	−1215	68.87	−1162	H(g)	218.0	114.6	203.3
CaH_2(s)	−189	42	−150	H_2(g)	0	130.6	0
CaO(s)	−635.5	40	−604.2	$H_2O(\ell)$	−285.8	69.91	−237.2
CaS(s)	−482.4	56.5	−477.4	H_2O(g)	−241.8	188.7	−228.6
$Ca(OH)_2$(s)	−986.6	76.1	−896.8	$H_2O_2(\ell)$	−187.8	109.6	−120.4
$Ca(OH)_2$(aq)	−1002.8	76.15	−867.6	*Iodine*			
$CaSO_4$(s)	−1433	107	−1320	I(g)	106.6	180.66	70.16
Carbon				I_2(s)	0	116.1	0
C(s, graphite)	0	5.740	0	I_2(g)	62.44	260.6	19.36
C(s, diamond)	1.897	2.38	2.900	ICl(g)	17.78	247.4	−5.52
C(g)	716.7	158.0	671.3	HI(g)	26.5	206.5	1.72
$CCl_4(\ell)$	−135.4	216.4	−65.27	*Iron*			
CCl_4(g)	−103	309.7	−60.63	Fe(s)	0	27.3	0
$CHCl_3(\ell)$	−134.5	202	−73.72	FeO(s)	−272	—	—
$CHCl_3$(g)	−103.1	295.6	−70.37	Fe_2O_3(s, hematite)	−824.2	87.40	−742.2
CH_4(g)	−74.81	186.2	−50.75	Fe_3O_4(s, magnetite)	−1118	146	−1015
C_2H_2(g)	226.7	200.8	209.2	FeS_2(s)	−177.5	122.2	−166.7
C_2H_4(g)	52.26	219.5	68.12	$Fe(CO)_5(\ell)$	−774.0	338	−705.4
C_2H_6(g)	−84.86	229.5	−32.9	$Fe(CO)_5$(g)	−733.8	445.2	−697.3
C_3H_8(g)	−103.8	269.9	−23.49	*Lead*			
$C_6H_6(\ell)$	49.03	172.8	124.5	Pb(s)	0	64.81	0
$C_8H_{18}(\ell)$	−268.8	—	—	$PbCl_2$(s)	−359.4	136	−314.1
$C_2H_5OH(\ell)$	−277.7	161	−174.9	PbO(s, yellow)	−217.3	68.70	−187.9
C_2H_5OH(g)	−235.1	282.6	−168.6	$Pb(OH)_2$(s)	−515.9	88	−420.9
CO(g)	−110.5	197.6	−137.2	PbS(s)	−100.4	91.2	−98.7
CO_2(g)	−393.5	213.6	−394.4				
CS_2(g)	117.4	237.7	67.15				
$COCl_2$(g)	−223.0	289.2	−210.5				

SELECTED THERMODYNAMIC VALUES AT 298.15 K *(continued)*

Species	ΔH_f^0 (kJ/mol)	S^0 (J/mol·K)	ΔG_f^0 (kJ/mol)	Species	ΔH_f^0 (kJ/mol)	S^0 (J/mol·K)	ΔG_f^0 (kJ/mol)
Lithium				*Rubidium*			
Li(s)	0	28.0	0	Rb(s)	0	76.78	0
LiOH(s)	−487.23	50	−443.9	RbOH(aq)	−481.16	110.75	−441.24
LiOH(aq)	−508.4	4	−451.1	*Silicon*			
Magnesium				Si(s)	0	18.8	0
Mg(s)	0	32.5	0	$SiBr_4(\ell)$	−457.3	277.8	−443.9
$MgCl_2(s)$	−641.8	89.5	−592.3	SiC(s)	−65.3	16.6	−62.8
MgO(s)	−601.8	27	−569.6	$SiCl_4(g)$	−657.0	330.6	−617.0
$Mg(OH)_2(s)$	−924.7	63.14	−833.7	$SiH_4(g)$	34.3	204.5	56.9
MgS(s)	−347	—	—	$SiF_4(g)$	−1615	282.4	−1573
Mercury				$SiI_4(g)$	−132	—	—
$Hg(\ell)$	0	76.02	0	$SiO_2(s)$	−910.9	41.84	−856.7
$HgCl_2(s)$	−224	146	−179	$H_2SiO_3(s)$	−1189	134	−1092
HgO(s, red)	−90.83	70.29	−58.56	$Na_2SiO_3(s)$	−1079	—	—
HgS(s, red)	−58.2	82.4	−50.6	$H_2SiF_6(aq)$	−2331	—	—
Nickel				*Silver*			
Ni(s)	0	30.1	0	Ag(s)	0	42.55	0
$Ni(CO)_4(g)$	−602.9	410.4	−587.3	*Sodium*			
NiO(s)	−244	38.6	−216	Na(s)	0	51.0	0
Nitrogen				Na(g)	108.7	153.6	78.11
$N_2(g)$	0	191.5	0	$Na^+(g)$	601	—	—
N(g)	472.704	153.19	455.579	NaBr(s)	−359.9	—	—
$NH_3(g)$	−46.11	192.3	−16.5	NaCl(s)	−411.0	72.38	−384
$N_2H_4(\ell)$	50.63	121.2	149.2	NaCl(aq)	−407.1	115.5	−393.0
$(NH_4)_3AsO_4(aq)$	−1268	—	—	$Na_2CO_3(s)$	−1131	136	−1048
$NH_4Cl(s)$	−314.4	94.6	−201.5	NaOH(s)	−426.7	—	—
$NH_4Cl(aq)$	−300.2	—	—	NaOH(aq)	−469.6	49.8	−419.2
$NH_4I(s)$	−201.4	117	−113	*Sulfur*			
$NH_4NO_3(s)$	−365.6	151.1	−184.0	S(s, rhombic)	0	31.8	0
NO(g)	90.25	210.7	86.57	S(g)	278.8	167.8	238.3
$NO_2(g)$	33.2	240.0	51.30	$S_2Cl_2(g)$	−18	331	−31.8
$N_2O(g)$	82.05	219.7	104.2	$SF_6(g)$	−1209	291.7	−1105
$N_2O_4(g)$	9.16	304.2	97.82	$H_2S(g)$	−20.6	205.7	−33.6
$N_2O_5(g)$	11	356	115	$SO_2(g)$	−296.8	248.1	−300.2
$N_2O_5(s)$	−43.1	178	114	$SO_3(g)$	−395.6	256.6	−371.1
NOCl(g)	52.59	264	66.36	$SOCl_2(\ell)$	−206	—	—
$HNO_3(\ell)$	−174.1	155.6	−80.79	$SO_2Cl_2(\ell)$	−389	—	—
$HNO_3(g)$	−135.1	266.2	−74.77	$H_2SO_4(\ell)$	−814.0	156.9	−690.1
$HNO_3(aq)$	−206.6	146	−110.5	$H_2SO_4(aq)$	−907.5	17	−742.0
Oxygen				*Tin*			
O(g)	249.2	161.0	231.8	Sn(s, white)	0	51.55	0
$O_2(g)$	0	205.0	0	Sn(s, grey)	−2.09	44.1	0.13
$O_3(g)$	143	238.8	163	$SnCl_2(s)$	−350	—	—
$OF_2(g)$	23	246.6	41	$SnCl_4(\ell)$	−511.3	258.6	−440.2
Phosphorus				$SnCl_4(g)$	−471.5	366	−432.2
P(g)	314.6	163.1	278.3	$SnO_2(s)$	−580.7	52.3	−519.7
P_4(s, white)	0	177	0	*Titanium*			
P_4(s, red)	−73.6	91.2	−48.5	$TiCl_4(\ell)$	−804.2	252.3	−737.2
$PCl_3(g)$	−306.4	311.7	−286.3	$TiCl_4(g)$	−763.2	354.8	−726.8
$PCl_5(g)$	−398.9	353	−324.6	*Tungsten*			
$PH_3(g)$	5.4	210.1	13	W(s)	0	32.6	0
$P_4O_{10}(s)$	−2984	228.9	−2698	$WO_3(s)$	−842.9	75.90	−764.1
$H_3PO_4(s)$	−1281	110.5	−1119	*Zinc*			
Potassium				ZnO(s)	−348.3	43.64	−318.3
K(s)	0	63.6	0	ZnS(s)	−205.6	57.7	−201.3
KCl(s)	−436.5	82.6	−408.8				
$KClO_3(s)$	−391.2	143.1	−289.9				
KI(s)	−327.9	106.4	−323.0				
KOH(s)	−424.7	78.91	−378.9				
KOH(aq)	−481.2	92.0	−439.6				

APPENDIX L

ANSWERS TO SELECTED EVEN-NUMBERED NUMERICAL EXERCISES

When calculations are done by different methods, roundoff errors may give slightly different answers. These differences are larger in calculations with several steps. Usually there is no cause for concern when your answers differ *slightly* from the answers given here.

CHAPTER 29

6. (a) 353 cm **(b)** 438

CHAPTER 30

No numerical exercises in this chapter.

CHAPTER 31

2. 0.024 mmol Bi/mL, 0.079 mmol Cu/mL, 0.045 mmol Cd/mL
6. (a) $1.1 \times 10^{-24} M$ **(b)** $1.1 \times 10^{-22} M$ **(c)** $1.1 \times 10^{-20} M$
8. $[Cu^{2+}] = 1.5 \times 10^{-16} M$; $[Pb^{2+}] = 1.4 \times 10^{-8} M$
40. $[Cu^{2+}] = 4.0 \times 10^{-15} M$; $[Bi^{3+}] = 1.2 \times 10^{-5} M$; $[Cd^{2+}] = 1.6 \times 10^{-8} M$

CHAPTER 32

2. 0.067 mmol As/mL, 0.041 mmol Sb/mL, 0.042 mmol Sn/mL

CHAPTER 33

2. $[Cr^{3+}] = 0.096 M$; $[Mn^{3+}] = 0.091 M$; $[Fe^{3+}] = 0.089 M$; $[Co^{2+}] = 0.085 M$; $[Ni^{2+}] = 0.085 M$; $[Zn^{2+}] = 0.077 M$; $[Al^{3+}] = 0.37 M$
76. (a) pH = 7.98 **(b)** pH = 3.53

CHAPTER 34

26. (b) $K' = 2.4 \times 10^{-15}$ **(c)** $Cr_2O_7^{2-}$
(d) $Cr_2O_7^{2-}$ at pH = 2.00, CrO_4^{2-} at pH = 12.00

CHAPTER 35

No numerical exercises in this chapter.

CHAPTER 36

8. $[Au^{3+}] = 1.4 \times 10^{-22} M$
12. $[Ag^+] = 2.1 \times 10^{-7} M$; $Q_{sp} < K_{sp}$, so Ag_2SO_4 will not precipitate
14. (a) 5.9×10^3 mol HBr (impossible)
(b) $(0.025 + 3.75 \times 10^{-5})$ mol ≈ 0.025 mol NaCN
(c) 0.060 mol $Na_2S_2O_3$ **(d)** $(0.050 + 4.4 \times 10^{-6})$ mol $\approx$ 0.050 mol NaCN **(e)** 0.99 mol NaCl
18. (a) HgS: $Q_{sp} = 5.8 \times 10^{-21} > 3.0 \times 10^{-53} = K_{sp}$, so HgS precipitates.
(b) CdS: $Q_{sp} = 5.8 \times 10^{-21} > 3.6 \times 10^{-29} = K_{sp}$, so CdS precipitates.
(c) ZnS: $Q_{sp} = 5.8 \times 10^{-21} > 1.1 \times 10^{-21} = K_{sp}$, so ZnS precipitates, but probably not enough to be seen.
(d) MnS: $Q_{sp} = 5.8 \times 10^{-21} < 5.1 \times 10^{-15} = K_{sp}$, so MnS does not precipitate.
20. (a) $[Hg^{2+}] = 1.5 \times 10^{-27} M$ **(b)** $[Cd^{2+}] = 1.8 \times 10^{-3} M$
(c) $[Hg^{2+}] = 5.5 \times 10^{-27} M$ and $[Cd^{2+}] = 6.5 \times 10^{-3} M$
22. $[Cd^{2+}] = 2.0 \times 10^{-6} M$, $[Pd^{2+}] = 7.7 \times 10^{-16} M$
24. (a) no **(b)** 0.44 M
26. (a) AgI **(b)** 99.95%

GLOSSARY/INDEX

Glossary terms are printed in **boldface** (with location of text definition indicated by boldface page numbers). Page numbers followed by *i* indicate illustrations or their captions; page numbers followed by *t* indicate tables.

USEFUL CONSTANTS

(For a more complete list, see Appendix D)

Atomic mass unit $1 \text{ amu} = 1.6606 \times 10^{-24} \text{ g}$

Avogadro's number $N = 6.0221367 \times 10^{23} \text{ particles/mol}$

Electronic charge $e = 1.60218 \times 10^{-19} \text{ coulombs}$

Faraday constant $F = 96{,}485 \text{ coulombs/equivalent}$
     $= 96{,}485 \text{ coulombs/mol } e^-$

Gas constant $R = 0.08206 \dfrac{\text{L} \cdot \text{atm}}{\text{mol} \cdot \text{K}} = 1.987 \dfrac{\text{cal}}{\text{mol} \cdot \text{K}}$
     $= 8.3145 \dfrac{\text{J}}{\text{mol} \cdot \text{K}} = 8.3145 \dfrac{\text{kPa} \cdot \text{dm}^3}{\text{mol} \cdot \text{K}}$

Ion product for water $K_w = 1.0 \times 10^{-14}$

Pi $\pi = 3.1416$

Planck's constant $h = 6.6262 \times 10^{-34} \text{ J} \cdot \text{s}$
     $= 6.6262 \times 10^{-27} \text{ erg} \cdot \text{s}$

Speed of light (in vacuum) $c = 2.99792458 \times 10^8 \text{ m/s}$

USEFUL RELATIONSHIPS

(For a more complete list, see Appendix C)

Mass and Weight

SI Base Unit: Kilogram (kg)

1 kilogram = 1000 grams = 2.205 pounds
1 gram = 1000 milligrams
1 pound = 453.59 grams
1 amu = 1.6606×10^{-24} grams
1 gram = 6.022×10^{23} amu
1 ton = 2000 pounds

Length

Si Base Unit: Meter (m)

1 inch = 2.54 centimeters (exactly)
1 meter = 100 centimeters = 39.37 inches
1 yard = 0.9144 meter
1 mile = 1.609 kilometers
1 kilometer = 1000 meters = 0.6215 mile
1 Ångstrom = 1.0×10^{-10} meters = 1.0×10^{-8} centimeters

Volume

SI Base Unit: Cubic Meter (m^3)

1 liter = 0.001 cubic meter
1 liter = 1000 cubic centimeters = 1000 mL
1 liter = 1.056 quarts
1 quart = 0.9463 liter
1 milliliter = 0.001 liter = 1 cubic centimeter
1 cubic foot = 7.475 gallons = 28.316 liters
1 gallon = 4 quarts

Energy

SI Base Unit: Joule (J)

1 calorie = 4.184 joules = 4.129×10^{-2} L · atm
1 joule = $1 \dfrac{\text{kg} \cdot \text{m}^2}{\text{s}^2} = 0.23901$ calorie
1 joule = 1×10^7 ergs
1 electron volt = 1.6022×10^{-19} joule
1 electron volt = 96.485 kJ/mol
1 L · atm = 24.217 calories = 101.325 joules

Pressure

SI Base Unit: Pascal (Pa)

1 pascal = $1 \dfrac{\text{kg}}{\text{m} \cdot \text{s}^2}$ = 1 newton/m^2
1 atmosphere = 760 torr
     = 760 millimeters of mercury
     = 1.01325×10^5 pascals
     = 1.01325 bar
     = 14.70 pounds per square inch
1 torr = 1 millimeter of mercury

Temperature

SI Base Unit: Kelvin (K)

$0 \text{ K} = -273.15°\text{C}$
$?\text{K} = °\text{C} + 273.15°$
$?°\text{F} = 1.8(°\text{C}) + 32°$
$?°\text{C} = \dfrac{°\text{F} - 32°}{1.8}$